Scanning Electron Microscopy

Scanning Electron Microscopy

Edited by **Lisa Page**

New York

Published by NY Research Press,
23 West, 55th Street, Suite 816,
New York, NY 10019, USA
www.nyresearchpress.com

Scanning Electron Microscopy
Edited by Lisa Page

International Standard Book Number: 978-1-63238-406-5 (Hardback)

Printed in the United States of America.

Contents

Preface

Over the recent decade, advancements and applications have progressed exponentially. This has led to the increased interest in this field and projects are being conducted to enhance knowledge. The main objective of this book is to present some of the critical challenges and provide insights into possible solutions. This book will answer the varied questions that arise in the field and also provide an increased scope for furthering studies.

Fine focused electron and ion beams constitute(s) an inevitable part of methods and instruments employed in various science fields. SEMs are well instrumented and supplemented with advanced techniques and methods and thereby present endless possibilities in the areas of quantitative measurement of object topologies, surface imaging, performing elemental analysis and local electrophysical characteristics of semiconductor structures. Creation of micro and nanostructures involves extensive use of fine focused e-beam. This book focuses on various issues concerned with scanning electron microscopy, covering both theoretical and practical aspects. Numerous topics are organized under two sections, "Material Science" and "Nanostructured Materials for Electronic Industry". This book includes contributions by renowned researchers and experts in this field.

I hope that this book, with its visionary approach, will be a valuable addition and will promote interest among readers. Each of the authors has provided their extraordinary competence in their specific fields by providing different perspectives as they come from diverse nations and regions. I thank them for their contributions.

Editor

Part 1

Material Science

Multimodal Microscopy for Ore Characterization

Otávio da Fonseca Martins Gomes and Sidnei Paciornik
Centre for Mineral Technology – CETEM,
Dept. of Materials Engineering, Catholic University of Rio de Janeiro,
Brazil

1. Introduction

The recent developments in electronics and computing have brought a radical change to the microstructural characterization of materials. The integration of digital image acquisition and digital image analysis with microscope automation methods is giving rise to a rich set of techniques in the new field of Digital Microscopy (Paciornik & Maurício, 2004).

Modern microscopes of all kinds (optical, electron, scanning probe) are controlled by software and have digital image acquisition. This setup allows many integrated tasks to be run under the control of automation routines like, for instance, specimen scanning and automatic focusing. Additionally, some microscopes can be fully automated. Thus, it is possible to integrate specimen scanning, image acquisition and storage, processing, analysis and report generation in a single routine.

Besides the automation of routine tasks in the microscopes, Digital Microscopy really opens new possibilities for microstructural characterization. In this context, multimodal microscopy emerges as a promising trend. Multimodal microscopy aims at combining complementary types of information from a given sample in order to build a multidimensional data set. It generates multi-component images combining layers obtained from different microscopy modalities, or from the same microscope in diverse conditions. For instance, multimodal microscopy may consider different signals in scanning electron microscopy (SEM) and different contrast modes in optical microscopy. Sometimes, multimodal microscopy is also referred as co-site, correlative or collaborative microscopy.

The key step of a multimodal microscopy methodology is the registration between images from a given field and/or set of fields. Image registration is the process of overlaying two or more images of the same scene taken at different conditions or by different sensors. Actually, registration is a crucial procedure in all image analysis tasks in which the final information is obtained from the combination of various data sources. Typically, registration is required to combine or compare images in remote sensing and medical imaging applications.

Once the multimodal set of images is acquired and registered, image segmentation can be employed to discriminate phases, regions or objects of interest. Due to the nature of this problem, multidimensional pattern recognition techniques arise as potential methods for image segmentation. Then, after segmentation, one is able to measure size, shape, intensity, and position parameters, leading to the possibility of automatic characterization of microstructural features.

The present chapter presents a multimodal methodology that combines images obtained by reflected light microscopy (RLM) and SEM. The so-called RLM-SEM co-site microscopy (Gomes & Paciornik, 2008a, 2008b) was developed to solve some ore microscopy problems that cannot be solved by either RLM or SEM.

2. Ore microscopy

Ore microscopy is an essential tool for ore characterization. It was generally employed in its various modalities (stereoscopic, transmitted and reflected light, SEM, etc.) for mineral identification and quantification, and in the determination of mineral texture and liberation analysis. In certain conditions, ore microscopy is the single approach to access this kind of information. In the mining industry, it is extensively used to provide parameters to the Geometallurgy procedures for exploration, production planning, and processing plant design and optimization purposes.

Transmitted and reflected light microscopy, respectively for transparent and opaque minerals, are probably the most traditional techniques of mineralogical identification. During the last two centuries, diverse analytical methods based on various properties of minerals were developed and refined. Referring to reflected light microscopy, it is worth to mention properties such as reflectivity, colour, reflection pleochroism, internal reflections, hardness, preferential polishing, chemical reactivity, crystalline habit, and crystalline texture, among others. There are some classical text-books that cover both theoretical and practical aspects of ore microscopy such as Galopin & Henry (1972), Gribble & Hall (1992), Criddle & Stanley (1993), and Craig & Vaughan (1994).

However, these traditional methods generally require an expert mineralogist and only few of them can be applied in automated systems. Thus, optical microscopy was being left aside in favour of SEM in ore characterization methodologies. In fact, in the last decades, research and development of microscopy in Applied Mineralogy field were focused on SEM.

The SEM is a very versatile analytical instrument. It builds images through synchronization of the electron beam scanning and one of the many signals that come from the interaction between the electron beam and the specimen. Thus, the pixels present intensities proportional to the signal measured by one of the SEM detectors such as, for instance, back-scattered electrons (BSE) or secondary electrons (SE) detectors. If the SEM has a coupled energy dispersive X-ray spectrometer (EDS), it becomes even more versatile, and can also perform elemental chemical analysis with a resolution down to approximately 1 μm on the surface. This is the great advantage of SEM – a large variety of electron-specimen interactions can be used to form images and to provide information with different physical meanings (Reimer, 1998; Goldstein et al., 2003).

The most used signal for ore characterization is BSE that can furnish topography information and atomic number contrast. Nevertheless, if the specimen is plane, each pixel is proportional to the average atomic number of its corresponding region on the specimen surface (Jones, 1987). Therefore, BSE images of polished samples are indirectly compositional images, in which mineral phases can be correlated to characteristic intensities or grey levels that are proportional to their average atomic numbers. Figure 1 shows a BSE image of iron ore in which four phases can be recognized: the embedding resin (the black background), quartz (dark grey), goethite (the grey particle at centre), and hematite (white). Table 1 presents chemical formula, colour on RLM, and average atomic number of epoxy resin and some minerals.

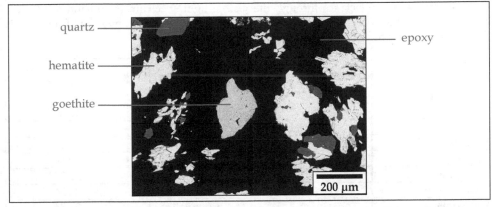

Fig. 1. BSE image of iron ore: the embedding resin (the black background), quartz (dark grey), goethite (the grey particle at centre), and hematite (white).

Based on BSE and EDS techniques, some automated systems for ore characterization were developed and commercially launched (Petruk, 1988; Sutherland & Gottlieb, 1991; Gu, 2003). These systems are SEM's especially dedicated to quantitative mineral analysis. They can identify minerals using BSE and EDS signals, and perform quantification routines through integrated image analysis software. Their capabilities may include particle-by-particle analysis, mineral phase classification and quantification, and mineral liberation analysis. Therefore, they became dominant for ore characterization, both in academy and industry, due to their enormous analytical capacity and relative simplicity of use.

Nevertheless, in recent years, there was a growing use of optical microscopy applied to ore characterization. Basically, three facts contributed to this trend: better optics, better digital image acquisition devices (Pirard et al., 1999), and the advent of Digital Microscopy. The progress in microscope optics, mainly due to infinity correct tubes and new advanced objective lenses, provided images with reduced spherical aberration and free of colour distortions (Davidson & Abramowitz, 1999), which are more suitable to image analysis and consequently to quantitative microscopy.

The colour has always been one of the most important properties used for mineral identification under a microscope (Piller, 1966). Moreover, there are some relevant minerals that are not distinguishable in the SEM, but can be discriminated through their colours in the reflected light microscope, such as, for instance, hematite and magnetite, which are the major iron ore minerals. Hematite and magnetite have similar average atomic numbers, respectively 20.59 and 21.02, and consequently show similar grey levels in BSE images, preventing their discrimination. The segmentation of hematite and magnetite in such kind of images requires a strong image contrast. However, this contrast condition avoids the segmentation of other present phases. In practice, not even SEM-based systems for automated mineralogy can discriminate hematite and magnetite, because the discrimination of these minerals is not possible through EDS due to their similar chemical composition (Gomes & Paciornik, 2008b).

On the other hand, transparent minerals and the embedding resin generally cannot be distinguished by their specular reflectances. For instance, quartz and epoxy resin have practically the same reflectance through the visible light spectrum (Neumann & Stanley, 2008). Actually, this is a classical problem in ore microscopy that renders unfeasible this kind of analysis through reflected light microscopy.

Phase	Chemical formula	Colour on RLM	Average atomic number
Epoxy resin	$C_{21}H_{25}ClO_5$	Dark grey	7.90
Quartz	SiO_2	Dark grey	10.80
Goethite	$FeO.OH$	Grey / Brown	19.23
Hematite	Fe_2O_3	Light grey	20.59
Pyrite	FeS_2	Pale yellow	20.66
Magnetite	Fe_3O_4	Pinkish grey	21.02
Pentlandite	$(Fe,Ni)_9S_8$	Pale yellow	23.36
Chalcopyrite	$CuFeS_2$	Brass yellow	23.54
Covelline	CuS	Blue	24.64
Bornite	Cu_5FeS_4	Purple	25.34
Sphalerite	ZnS	Grey	25.39
Chalcocite	Cu_2S	Light grey	26.38
Native copper	Cu	Bright yellow	29.00

Table 1. Chemical formula, colour on RLM, and average atomic number of epoxy resin and some minerals.

Figure 2 shows a pair of images of an iron ore sample acquired by reflected light microscopy and SEM. Comparing them, one can observe that the segmentation between quartz and epoxy resin in the BSE image is easy, but it is not viable in the optical image. On the other hand, hematite and magnetite present distinct colours, respectively light grey and pinkish grey, in the optical image, but have practically the same grey level in the BSE image.

Another example of minerals of difficult discrimination can be observed in Figure 3. It shows images of the same field of a copper ore sample acquired by reflected light microscopy and SEM. In the optical image, chalcopyrite can be easily identified by its characteristic brass yellow colour, but pyrite and pentlandite present a very similar colour (pale yellow). On the other hand, in the BSE image, chalcopyrite and pentlandite are practically indistinguishable, due to their similar average atomic numbers (23.54 and 23.36, respectively). Nevertheless, pyrite is slightly darker than pentlandite, because pyrite has a lower average atomic number (20.66).

The RLM-SEM co-site microscopy was developed to overcome these challenges. This methodology can improve the SEM analytical capacity adding specular reflectance (colour) information from RLM. The methodology was employed with some mineral samples, aiming at the discrimination of phases that are not distinguishable by either RLM or SEM, but can be discriminated through the combined use of both techniques.

Fig. 2. Images of an iron ore sample acquired on (a) reflected light microscope and (b) SEM.

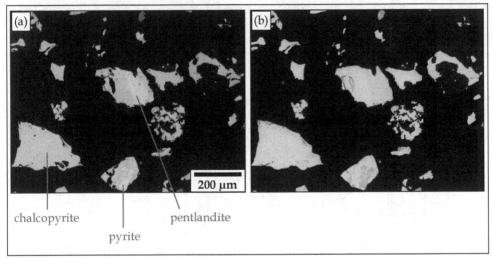

Fig. 3. Images of a copper ore sample acquired on (a) reflected light microscope and (b) SEM.

3. Image registration

Image registration comprises the operation to determine the correspondence point to point between two or more images of the same area (or volume) obtained by different sensors or in different conditions, and the subsequent process of overlaying them.

Image registration is a fundamental procedure in all image analysis tasks in which the final information is gained from the combination of various data sources. Only after the registration, a multi-component image that represents a multimodal database can be properly composed and analyzed.

Typically, image registration is employed for composition and comparison of multi-spectral images in Remote Sensing (Schowengerdt, 1983). It also has several applications in Medicine, such as diagnosis, preparation of surgeries, treatment evaluation, etc. It is used, for instance, for fusion of anatomical and functional information, which are usually obtained through different medical imaging techniques (van den Elsen et al., 1993; Maintz & Viergever, 1998).

There is in the literature a wide variety of image registration methods based on different principles and employed for diverse applications (Zitova & Flusser, 2003; Goshtasby, 2005). Anyway, registration consists in the determination and implementation of a geometric operation (spatial transformation) between images in order to correlate the spatial coordinates of both images. Therefore, the fundamental aspect of any registration method is the spatial transformation used to correctly overlay images. Although many types of variations may be present in images, a suitable transformation must remove only spatial distortions between them (Brown, 1992). Other differences, due to the diversity of information that each image represents, must be maintained, since these are the interesting characteristics that one aims to expose.

In fact, image registration is more complex than a simple image alignment. It is not limited to translation and rotation of images. It may be composed of a combination of six distinct basic transformations: translation, rotation, scale, shear, projection, and other non-linear and local distortions. Figure 4 presents the six basic transformations, showing their effects in a sample base image.

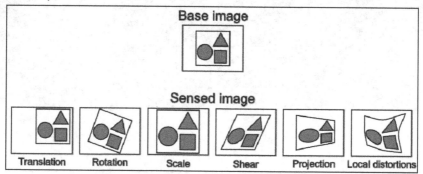

Fig. 4. Basic spatial transformations.

Spatial transformations can convert the coordinates of the sensed image to the coordinates of the base or reference image. Thus, they correlate these digital images pixel by pixel, allowing the assemblage of a multi-component image.

The transformations that involve only translation and rotation are generally called rigid body or Euclidean transformations, since the Euclidean distances within images are preserved (Szeliski, 2004). In contrast, the other ones are classified as non-rigid or elastic. Although this nomenclature is the most commonly found in the literature, including the present text, it is not a consensus. Some authors consider scale as a rigid body transformation too, and there are still others that also include shear and projection in the class of rigid transformations (Crum, et al., 2004).

In this context, multimodal microscopy procedures that are performed intrinsically in a unique microscope constitute probably the simplest cases, generally involving only rigid body transformations. Sometimes it is even possible to acquire images that are directly registered.

Multimodal microscopy methodologies on reflected light microscope can be carried out with or without specimen removal from the stage. In the first case, specimen removal, for instance, for chemical etchings like in a classical metallographic approach, generally imply some displacement between images. Therefore, translation and occasionally rotation corrections are required. Soto et al. (2004) and Paciornik & Gomes (2009) present case studies of multimodal methods that involve specimen removal for chemical etchings.

On the other hand, optical methodologies without specimen removal can be sometimes performed without translation corrections (De-Deus et al., 2007). Nevertheless, there are some exceptions. For instance, Pirard (2004) proposed a multispectral imaging technique applied to ore characterization in which shifts of the order of several pixels occur between images obtained from different wavelengths; and Iglesias et al. (2011) developed a multimodal microscopy methodology based on the combination of cross-polarized and bright field images in which there were small misalignments between them.

The SEM forms an image through scanning its electron beam in a raster across the specimen and then it synchronizes the scanning with a signal from one of its detectors. Thus, in a given field, it can acquire several different images, which are ready to compose a multi-component image without the need for a registration procedure. However, in practice, older equipments usually exhibit some translation between images from different detectors. In this case, a translation correction is not enough to properly register the images, because SEM's generally present complex and non-linear distortions (Goldstein et al., 2003) that must be considered.

In the RLM-SEM co-site microscopy, the registration procedure involves rigid and non-rigid transformations. The specimen handling between the microscopes and the different stages imply that translation and rotation adjustments are necessary. Besides, non-rigid transformations are required due to the complex distortions that occur in images from SEM. Even a fine calibration of the equipment is not capable of preventing them.

A registration procedure consists of a sequence of mathematical operations that determine the suitable spatial transformation and then defines and applies the geometric operation that properly performs the registration. The base of a registration procedure is the kind of information used by its algorithm. Therefore, as stated in the already classical review paper of Zitova & Flusser (2003), there are two main classes of algorithms according to their nature: area-based and feature-based.

Area-based algorithms, also called template matching, estimate the correspondence between images (or parts of them) in order to determine which transformations provide the best correspondence. The correlation between two signals (cross-correlation) is a standard approach to template matching algorithms that can be particularly efficient if it is computed in the frequency domain using the fast Fourier transform (Lewis, 1995). Area-based algorithms are in general simpler than feature-based ones. They are applied mostly in cases involving only rigid and scale transformations. Besides, they are more sensitive to noise in images. For instance, the multimodal methodologies presented by Soto et al. (2004), Paciornik & Gomes (2009), and Iglesias et al. (2011) employed cross-correlation in the frequency domain for the registration of their optical images.

Feature-based algorithms consist of four steps: feature detection, feature matching, mapping function design, and image transformation and resampling (Zitova & Flusser, 2003). Two sets of features, which are salient and distinctive objects such as corners, line intersections, edges, etc., are manually or automatically detected in both base and sensed images. These features are represented by the so-called control points (points themselves, centers of gravity, line endings, etc.). The aim is to find the pairwise correspondence between control points and then to map a suitable transformation from them. Therefore, the sensed image is transformed through the determined mapping function and an appropriate interpolation technique is employed in order to treat non-integer coordinates.

The detection of control points and the determination of their correspondence in base and sensed images are crucial and difficult tasks. The method named Scale Invariant Feature Transform (SIFT), proposed by Lowe (2004), has been shown computationally efficient and robust upon diverse distortions and multisensor cases.

In contrast to the area-based methods, the feature-based ones do not work directly with image intensity values. Control points constitute higher level information. This fact makes feature-based methods suitable for applications in which diverse sensors with different data structures and physical meanings are involved. Besides, it allows registering images with any nature of distortions, including non-linear and local ones (Zitova & Flusser, 2003).

Furthermore, in multimodal microscopy methodologies, an alternative approach can facilitate the determination of control points. By introducing indentation marks in the sample through a microdurimeter, the control points can be properly defined as their centers of gravity. In fact, this method can be useful for the registration of one or few fields in which specific microstructural features are of interest. However, it becomes impractical when the number of fields is large, as usually occurs in ore characterization procedures.

The RLM-SEM co-site microscopy methodology employs a feature-based method for registration that is described in the section 5.3.

4. Image processing and analysis

A typical image processing and analysis sequence comprises the steps of image acquisition, digitization, pre-processing, segmentation, post-processing, feature extraction and classification (Gonzalez & Woods, 2007).

Pre-processing, or image enhancement, is the first step after image digitization and is used to correct basic image defects, normally created during the image acquisition step. Typical operations, at this step, are background correction, edge enhancement and noise reduction. Pre-processing is useful for qualitative reasons, as it improves the visibility of relevant features in the image, but it is even more important to prepare the image for the following step of segmentation.

Segmentation is the technical term used for the discrimination of objects in an image. Segmentation is probably the most complex step in the sequence because it tries to represent computationally a cognitive process that is inherent to the human vision. When we look at an image we use many different inputs to distinguish the objects: brightness, boundaries, specific shapes or textures. Our brains process this information in parallel at high speed, using previous experience. Computers, on the other hand, do not have the same associative power. The recognition of objects in an image is made through the classification of each pixel of the image as pertaining or not to an object.

There are many segmentation methods based on different principles such as thresholding, edge detection, texture analysis, mathematical morphology, etc. Each one is generally more suitable for a specific application. Categorically, there is not an ideal generic method that is always the best one. Some classical references in this area are: Haralick et al. (1973), Otsu (1979), Haralick (1979), Beucher & Lantuéjoul (1979), Marr & Hildreth (1980), Pun (1981), Canny (1986), and Adams & Bischof (1994).

The most common segmentation method is thresholding. It is based on the assumption that pixels pertaining to a given class of objects (e.g. a specific mineral phase) have similar colour or greyscale intensity, and this colour is different from the background and from other classes of objects. In other words, there must be sufficient contrast between different phases in the material. If that is the case, then the segmentation is based on selecting colour/intensity thresholds that represent the various phases.

Noise, uneven illumination, edge effects contribute to degrade the discrimination between phases, and that is why a pre-processing step may be so relevant. Evidently, phase contrast maybe too low, depending on the microscope used, as described before, and that is where combining information from different types of signals becomes critical.

In many situations the results of segmentation contain artefacts, such as spurious objects, touching or partially overlapping objects, etc. A very common artefact in mineralogical images is segmenting a phase together with the edges of a different phase that share the same intensity range. Some of these defects can be minimized with an appropriate pre-processing step, such as delineation (edge enhancement), but many must be corrected after segmentation, in the so-called post-processing step.

Post-processing makes intense use of morphological operators such as erosion, dilation, opening, closing, and more sophisticated functions like the watershed separation method for touching objects (Serra, 1982, 1988). Ideally, the final result is an image in which just the relevant objects are present and separated in groups that correspond to each phase present in the sample. However, as described below, this is rarely the case, and further analysis of the segmented objects must be done to complete the discrimination.

Given a segmented post-processed image containing a set of objects, several measurements are available. Field features such as number of objects and area fraction are some of the simplest ones. Object specific features are more sophisticated and include measurements of size, shape, position, intensity and texture of each object in the image (Friel, 2000). These features are critical for the classification step.

4.1 Multi-component image analysis

Common colour images, generated by either scientific or general-purpose digital cameras, are generally 24-bit RGB images. Actually, they consist of multi-component images composed by three images that represent the primary colours (red, green and blue) with 8-bit quantization each (Orchard & Bouman, 1991). The RGB system is the most common colour representation system employed by cameras and displays. It was developed to be similar with the *tri-stimulus* response of human vision. Figure 5 shows a sample RGB image above its three components, respectively, R, G and B. In this figure, one can observe some samples of how the primary colours are mixed to compose other colours and grey levels.

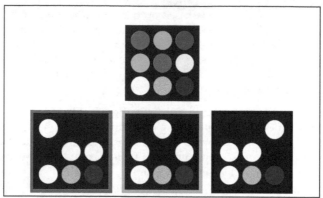

Fig. 5. A RGB image and its components.

The components of a multi-component image can represent information of any source and physical meaning. It is not necessary that the data within components are correlated. They just must have spatial correspondence pixel-to-pixel, i.e., they must be registered.

Multi-component images with up to three components have the advantage that they can be viewed as RGB colour images in standard image viewer software. In this case, the so-called pseudo-colours denote the properties which are represented in the components. Pseudo-colour images constitute a useful approach for data visualization.

Each component image can be singly processed and analyzed as a common intensity or grey level image. However, this processing should be carefully performed so that the spatial relations within images are not undesirably modified. Geometric operations must be especially avoided.

A multi-component image can also be conceived as a matrix in which each element is a vector, not a scalar value. Each vector represents a pixel, and each element of vector is the value of this pixel in one of the component images. In a RGB image, each pixel consists of a three-element vector that represents the intensity of the three primary colours. Therefore, in a multi-component image, the probability density function becomes multivariate and consequently its histogram of intensities becomes multidimensional. Figure 6 shows a RGB image of a copper ore sample in the reflected light microscope and its bi-dimensional RG histogram.

Image segmentation by thresholding can be generalized to an n-dimension problem. In this case, a threshold becomes a decision boundary, whose form depends on the number of components. One component leads to decision boundaries that are scalar values (thresholds); two components imply that decision boundaries are straight lines; three components involve decision boundaries that are planes; and so on. Nevertheless, there are more sophisticated segmentation methods that offer more complex decision boundaries, such as curves, surfaces, etc. Besides, they can also be discontinuous. In this context, multidimensional pattern recognition techniques arise as potential methods for segmentation of multi-component images.

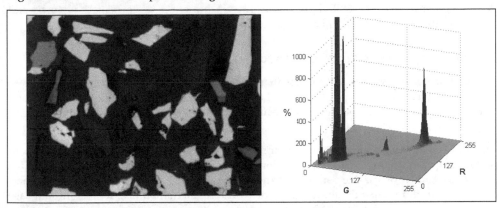

Fig. 6. A RGB image and its bi-dimensional RG histogram.

Pattern recognition is the scientific discipline whose goal is the classification of objects (patterns) into a number of classes from the observation of their characteristics (Theodoridis & Koutroumbas, 2003). It aims to build a simpler representation of a data set through its more relevant features in order to perform its partition into classes (Duda et al., 2001).

Pattern recognition techniques can be used to classify objects (pixels, regions, etc.) within images. Any part of image that has at least one measurable property can be considered as an object and consequently it can be classified. In multi-component images, a pixel consists of a vector in which the elements represent its values in the components.

The classification of pixels is actually an image segmentation procedure. Each class of pixels can properly represent a phase or mineral in an ore microscopy case. If the phases of interest are known, this problem becomes supervised (Duda et al., 2001). In a supervised classification procedure, this known information is exploited so that the classification system learns how the different classes can be recognized.

A supervised classification procedure involves three stages: training, validation and classification. The training stage comprises sampling of known pixels of each class in order to compose the so-called training set that is used as knowledge base. Therefore, the classifier is trained, i.e., it is designed based on the known information. Following, in the validation stage, another known set of pixels, the validation set, is classified aiming to estimate the performance of the classifier, considering its generalization capacity (Toussaint, 1974). If the validation stage indicates that the training was successful, the classification is then possible and consequently the segmentation procedure is implemented.

It is worth to mention that although the RGB system is vastly predominant in image acquisition devices, it is not generally the most appropriate colour representation system for classification purposes because its three components (R, G and B) are very correlated, due to their strong dependence from the light intensity (Littmann & Ritter, 1997). Besides, it doesn't represent colours in a uniform scale, preventing measurements of similarity between colours through their distance in RGB space (Cheng et al., 2001). There are many colour systems described in the literature (Sharma & Trussell, 1997). Actually, one can define any colour system from linear or non-linear transformations of RGB (Vandenbroucke et al., 2003). Systems like rgb, HSI, L*a*b* and L*u*v* present less correlated features and consequently tend to be more suitable for classification (Gomes & Paciornik, 2008a).

5. Combining reflected light microscopy and SEM

This section describes the RLM-SEM co-site microscopy methodology by reviewing two case studies, in which it was applied for the characterization of a copper ore and an iron ore. These case studies were originally presented by Gomes & Paciornik (2008a and 2008c, respectively).

The RLM-SEM co-site microscopy comprises four sequential steps: image acquisition in RLM; image acquisition in SEM; registration; and image processing and analysis. The three first steps consist of generic routines of the methodology, but the image processing and analysis procedure depends on the case study. The segmentation of minerals in both case studies was performed through supervised classification of pixels, exploiting their multidimensional nature. However, used features and classification methods differ.

Ore microscopy procedures commonly involve acquisition and analysis of tens to hundreds of images per cross-section in order to provide a representative sampling. Therefore, in the development of the RLM-SEM co-site microscopy, automatic routines for field scanning and image acquisition were implemented for both used microscopes.

5.1 Image acquisition in reflected light microscopy

A motorized and computer controlled microscope with a digital camera (1300 x 1030 pixels) was used. A routine was implemented for microscope and camera control, and for image acquisition. It integrates and automates many common procedures such as specimen x-y scanning, automatic focusing, background correction and imaging.

The following image acquisition procedures and conditions were employed:

a. Before image acquisition, a SiC reflectivity standard was used to generate background images for each objective lens, which were subsequently employed for automatic background correction (Pirard et al., 1999) of every acquired image.
b. Illumination was kept constant by direct digital control of the lamp voltage.
c. Camera sensitivity, exposure and white balance were optimized initially for a representative field of the sample and kept constant there on.
d. Objective Lenses: 5X (NA 0.13); 10X (NA 0.20); 20X (NA 0.40), leading to resolutions of 2.11, 1.05, and 0.53 µm/pixel, respectively.
e. Single fields regularly spaced on the sample were imaged through specimen scanning with a motorised stage and automatic focusing.
f. Each field position was recorded in a database for subsequent image acquisition on in SEM.
g. All images were acquired at 24 bit RGB colour quantisation.

5.2 Image acquisition in SEM

A digital SEM was used to acquire a BSE image (1024 x 768 pixels) of each field imaged on RLM. In this procedure, the sample must be placed in the SEM stage at a similar arrangement as positioned in the RLM stage. It is unnecessary and impractical to place the sample in the exact same way, but a similar arrangement can make image registration easier and faster.

The magnification of the SEM was set to keep similar resolutions to optical images. Other SEM operational parameters were manually tuned for a representative field of the sample and then kept constant. After that, the field positions database was loaded with the acquisition routine developed in the SEM control software. It converts RLM stage coordinates to SEM stage coordinates and subsequently performs automatic specimen scanning and image acquisition. Thus, respectively for copper ore and iron ore samples, 121 (11 x 11) and 81 (9 x 9) fields per cross-section were imaged with the RLM and the SEM. Figure 7 presents a field of the copper ore sample as imaged on RLM and SEM.

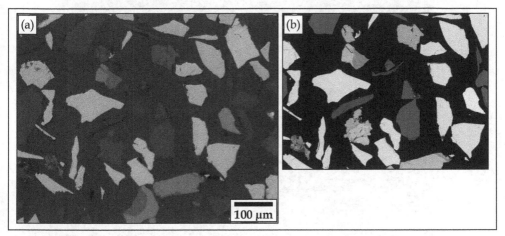

Fig. 7. Images of a field of the copper ore sample obtained by (a) RLM and (b) SEM.

5.3 Image registration

An automatic method for the registration of RLM and SEM images was developed. The distortions were considered according to their sources, and the registration procedure was accomplished through sequential steps. The first step comprises distortions from the SEM, such as astigmatism and local distortions. The second step adjusts rotation and the third one corrects translation. At the end, the registered images are cropped to represent exactly the same field.

The first step is carried out through a feature-based registration algorithm. It maps the SEM characteristic distortions based on several control points that are automatically detected. These distortions do not depend on samples. They are a function of SEM operational parameters, such as magnification and working distance. Therefore, this step was employed only one time for each SEM set-up, i.e., once for each magnification whose pixel size corresponds to the pixel size obtained through one of the objective lenses of the RLM.

A standard specimen with regular distributed and easily extractable control points must be imaged on RLM and SEM. In the present case, two copper grids (200 mesh for 5X and 10X lenses, and 400 mesh for 20X lens) were used. These images were analyzed by an automatic routine in order to determine the centroid (center of gravity) of each grid hole, the control points. Then, a suitable spatial transformation was computed from the pairs of control points using the local weighted mean method proposed by Goshtasby (1988). Therefore, this spatial transformation was applied to just remove distortions in every SEM image of the ore sample.

Figure 8 shows the pairs of control points superimposed to the RLM image of the grid (10X). The white circles represent the control points extracted from the RLM image, and the white dots are the points obtained from the corresponding SEM image. As can be observed, circles and points are not aligned. Besides, the misalignment varies, evidencing the complexity of these distortions.

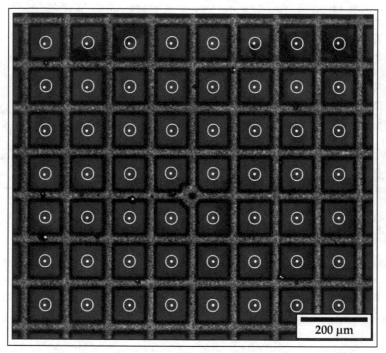

Fig. 8. Pairs of control points superimposed on the RLM image of the grid (10X). The white circles represent the control points extracted from the RLM image, and the white dots are the control points obtained from the corresponding SEM image.

The second step of the registration method aims at finding the rotation angle between images and subsequently to adjust rotation. This angle is due to sample manipulation and its different arrangement in the sample holders of the microscopes. Thus, this rotation is constant in all fields of a sample in a given experiment.

An iterative algorithm is used to determine the angle that maximizes the normalized cross correlation between a pair of images. The algorithm uses coarse-to-fine approach. It evolves in order to adjust the angle down to 0.01° of precision. This procedure is applied for one pair of images and the obtained angle is used to correct rotation in all SEM images.

After the second step, the SEM images are free of distortions and they are put in the same coordinate system of the RLM images. Therefore, only translation problems remain. Thus, in the third step, the RLM and the SEM images are finally registered through the maximization of normalized cross correlation. At the end, they are cropped to represent exactly the same field. Figure 9 shows the images present in Figure 7, after registration.

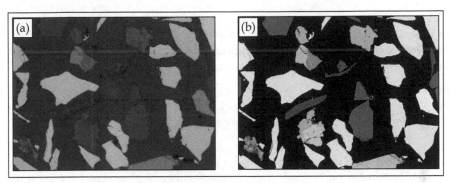

Fig. 9. (a) RLM and (b) SEM images from Figure 7, after registration.

5.4 Image processing and analysis

The registered SEM and RLM images went through a pre-processing step of delineation to reduce the well-known halo effect, making them more suitable for the subsequent segmentation procedures. Figure 10 shows a detail of an image obtained by RLM, before and after delineation operation. Comparing them, one can observe that delineation improves phase transitions and consequently allows better segmentation results.

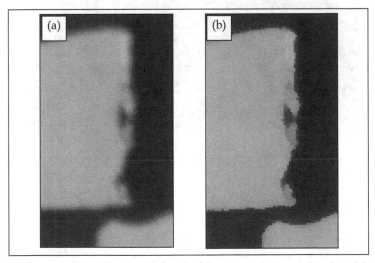

Fig. 10. Delineation of a RLM image: (a) detail of the original image; (b) after delineation.

5.4.1 Copper ore

The copper ore sample had a complex mineralogy mainly composed of thirteen minerals, which were then taken as individual classes (quartz, three different silicates, apatite, magnetite, pentlandite, chalcopyrite, covelline, bornite, sphalerite, chalcocite, and native copper). Besides mineral phases, epoxy resin was taken as a class too. Thus, the training stage involved sampling of pixels from the fourteen classes. In practice, 6000 pixels of each phase were interactively selected from several images.

The delineated RLM images were converted from RGB to the rgb, HSI, L*a*b* and L*u*v* colour systems with the goal of revealing colour information hidden by the correlated RGB system. These conversion operations generate ten new components (r, g, b, H, S, I, a*, b*, u*, v*), increasing the system dimensionality from four to fourteen.

A Bayes classifier (Duda et al., 2001) was employed and the fourteen components (BSE intensity and the thirteen colour components) were used as features. The validation was carried out through holdout estimate (Toussaint, 1974), reaching a success rate larger than 99.5%.

The classification result was a grey level image per field in which each phase was represented by a distinct grey level. Thus, pixels classified as phase one have intensity one, pixels recognised as phase two have intensity two, and so on. Therefore, a suitable look-up table can be applied in order to attribute a different pseudo-colour to each phase and consequently to make their visualisation easier. Figure 11 presents the classification result for the images shown in Figure 9 and its look-up table.

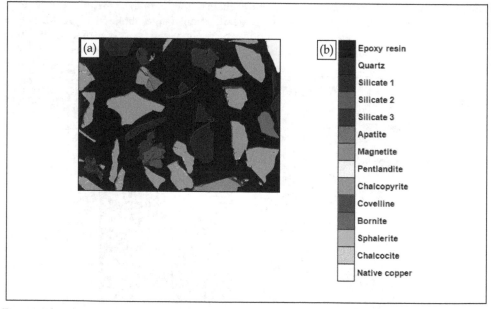

Fig. 11. The classification result for the images shown in the Figure 9: (a) segmented image; and (b) look-up table.

5.4.2 Iron ore

The mineralogical assemblage of the iron ore sample was simple. It was mainly composed by hematite, magnetite, goethite and quartz. Therefore, in this case study, five classes were considered (epoxy resin, quartz, goethite, haematite, and magnetite).

The segmentation process was split in two supervised classification procedures. The first one recognised epoxy resin, quartz, goethite, and a hematite-magnetite composed phase through the classification of pixels in SEM images, using the BSE intensity as feature in a Bayes classifier. Then, the second classification procedure was able to discriminate hematite and magnetite. It was carried out through the classification of pixels in RLM-SEM composed images. In this case, their four components (R, G, B, and BSE intensity) were used as features and a Bayes classifier was employed. The training stage for both classification procedures comprised interactive sampling of 1000 pixels of each one of the five classes from a RLM-SEM composed image. Figure 12 presents the segmentation result for the images shown in Figure 2 and its applied look-up table.

Fig. 12. The classification result for the images shown in the Figure 2: (a) segmented image; and (b) look-up table.

5.5 Discussion

The information from RLM and SEM presents different structures and physical meanings. RLM data consists of a vector of three 8-bit values, which denote specular reflectance in the RGB colour system. On the other hand, SEM data is represented as 8-bit values of BSE intensity that provides average atomic number contrast. Thus, it is very difficult to compute suitable measurements of similarity between patterns (pixels) that can be used to recognise the classes. In this kind of problem, Valev and Asaithambi (2001) pointed out that different classifiers can be used to complement one another. This is the approach employed for the segmentation in the iron ore case study.

The increase of dimensionality carried out in the copper ore case study can make the classification task easier, since hidden information is discovered and consequently patterns are better described. However, the training data must grow exponentially with the dimensionality in order to prevent the so-called curse of dimensionality (Marques de Sá,

2001). In practice, this is not an issue for pixels classification as a typical image of 1024 x 768 pixels, for instance, has about 0.8 million of pixels, and it is easy to obtain several thousands of pixels of each class.

The resulting segmented images in both case studies reveal small amounts of misclassified pixels in borders between phases. It occurs mainly due to little cracks and relief, in spite of the good sample preparation and the delineation pre-processing. This misclassification is quite small and it can be negligible in mineralogical identification and quantification procedures. However, it becomes more significant for microstructural characterisation purposes, such as mineral liberation analysis. Therefore, post-processing routines should be developed.

6. Conclusion

Multimodal microscopy extends the capabilities of traditional microscopy techniques, improving the discrimination of mineral phases in ores. By combining Optical Microscopy and Scanning Electron Microscopy it takes advantage of the complementary contrasts provided by these techniques.

This method relies on microscope automation, digital image acquisition, processing and analysis. Over the last years many of these techniques have become readily available in both commercial and free software environments.

The use of supervised classification methods relies on operator experience, during the training stage, but once the classifier is optimized and validated, the effective classification of unknown samples is fully automatic and fast.

The developed method is applicable to other materials for which individual microscopy techniques do not provide enough discrimination between the relevant phases.

7. Acknowledgment

One of the authors (S. Paciornik) acknowledges the support of CNPq, the Brazilian Research Council.

8. References

Adams, R. & Bischof, L. (1994). Seeded region growing. *IEEE Transactions on Pattern Analysis and Machine Intelligence*, Vol. 16, No. 6, (June 1994), pp. 641–647, ISSN 0162-8828

Beucher, S. & Lantuéjoul, C. (1979). Use of watersheds in contour detection, *Proceedings of International Workshop on Image Processing, Real-time Edge and Motion detection/estimation*, pp. 2.1–2.12, Rennes, France, September 17-21, 1979

Brown, L.G. (1992). A Survey of Image Registration Techniques. *ACM Computing Surveys*, Vol. 24, No. 4, (December 1992), pp. 325-376, ISSN 0360-0300

Canny, J. (1986). A computational approach to edge detection. *IEEE Transactions on Pattern Analysis and Machine Intelligence*, Vol. 8, No. 6, (November 1986), pp. 679–698, ISSN 0162-8828

Cheng, H.D.; Jiang, X.H.; Sun, Y. & Wang, J. (2001). Color image segmentation: advances and prospects. *Pattern Recognition*, Vol. 34, No. 12, (December 2001), pp. 2259-2281, ISSN 0031-3203

Criddle, A.J. & Stanley, C.J. (1993). *Quantitative Data File for Ore Minerals* (3rd Edition), Chapman & Hall, ISBN 041246750X, London, UK

Craig, J.R. & Vaughan, D.J. (1994). *Ore microscopy and ore petrography* (2nd Edition), John Wiley & Sons, ISBN 0471551759, New York, USA

Crum, W.R.; Hartkens, T. & Hill, D.L.G. (2004). Non-rigid image registration: theory and practice. *The British Journal of Radiology*, Vol. 77, (December 2004), pp. S140-S153, ISSN 0007-1285

Davidson, M.W. & Abramowitz, M. (1999). *Optical microscopy*, The Florida State University, Retrieved from: <http://micro.magnet.fsu.edu/primer/opticalmicroscopy.html>

De-Deus, G.; Reis, C.M.; Fidel, R.A.S.; Fidel, S.R.; Paciornik, S. (2007). Co-site digital optical microscopy and image analysis: an approach to evaluate the process of dentine demineralization. *International Endodontic Journal*, Vol. 40, No. 6, (June 2007), pp. 1365-2591, ISSN 1365-2591

Duda, R.O.; Hart, P.E. & Stork, D.G. (2001). *Pattern classification* (2nd Edition), Wiley-Interscience, ISBN 0-471-05669-3, New York, USA

Friel, J.J. (2000). Measurements, In *Practical guide to image analysis*, pp. 101-128, ASM International, ISBN 0871706881, Materials Park, USA

Galopin, R. & Henry, N.F.M. (1972). *Microscopic study of opaque minerals*, W. Heffer and Sons Ltd., ISBN 0852700474, Cambridge, UK

Goldstein, J.I.; Newbury, D.E.; Echlin, P.; Joy, D.C.; Lyman, C.E.; Fiori, C.; Lifshin, E.; Sawyer, L. & Michael, J.R. (2003). *Scanning Scanning Electron Microscopy and X-ray Microanalysis* (3rd Edition), Springer, ISBN 0306472929, New York, USA

Gomes, O.D.M. & Paciornik, S. (2008a). Co-site microscopy: combining reflected light and scanning electron microscopy to perform ore mineralogy, *ICAM 2008 - Ninth International Congress for Applied Mineralogy Conference Proceedings*, pp. 695-698, ISBN 9781920806866, Brisbane, Australia, September 8-10, 2008

Gomes, O.D.M. & Paciornik, S. (2008b). Iron ore quantitative characterization through reflected light-scanning electron co-site microscopy, *ICAM 2008 - Ninth International Congress for Applied Mineralogy Conference Proceedings*, pp. 699-702, ISBN 9781920806866, Brisbane, Australia, September 8-10, 2008

Gomes, O.D.M. & Paciornik, S. (2008c). RLM-SEM co-site microscopy applied to iron ore characterization, *3rd International Meeting on Ironmaking and 2nd International Symposium on Iron Ore Conference Proceedings*, pp. 218-224, ISBN 9788577370320, São Luís, Brazil, September 22-26, 2008

Gonzalez, R.C. & Woods, R.E. (2007). *Digital Image Processing* (3rd Edition), Prentice-Hall, ISBN 013168728X, Upper Saddle River, USA

Goshtasby, A. (1988). Image Registration by Local Approximation Methods. *Image and Vision Computing*, Vol. 6, No. 4, (November 1988), pp. 255-261, ISSN 0262-8856

Goshtasby, A. (2005). *2-D and 3-D image registration for medical, remote sensing, and industrial applications*, John Wiley & Sons, ISBN 0471649546, Hoboken, USA

Gribble, C. & Hall, A.J. (1992). *Optical Mineralogy: Principles and Practice*, UCL Press, ISBN 185728013X, London, UK

Gu, Y. (2003). Automoated Scanning Electron Microscope Based Mineral Liberation Analysis - An Introduction to JKMRC/FEI Mineral Liberation Analyser. *Journal of Minerals and Materials Characterisation and Engineering*, Vol. 2, No. 1, pp. 33-41, ISSN 1539-2511

Haralick, R.M.; Shanmugam, K.; Dinstein, I. (1973). Textural features for image classification. *IEEE Transactions on Systems, Man, and Cybernetics*, Vol. 3, No. 6, (November 1973) pp. 610-621, ISSN 0018-9472

Haralick, R.M. (1979). Statistical and Structural Approaches to Texture. *Proceedings of the IEEE*, Vol. 67, No. 5, (May 1979), pp. 786-808, ISSN 0018-9219

Iglesias, J.C.A.; Gomes, O.D.M. & Paciornik, S. (2011). Automatic recognition of hematite grains under polarized reflected light microscopy through image analysis. *Minerals Engineering*, Vol. 24, No. 12, (October 2011), pp. 1223-1378, ISSN 0892-6875

Jones, M.P. (1987). *Applied mineralogy: a quantitative approach*, Graham and Trotman Ltd., ISBN 0860105113, London, UK

Lewis, J.P. (1995). Fast Template Matching, *Vision Interface 95*, pp. 120-123, Quebec City, Canada, May 15-19, 1995

Littmann, E. & Ritter, H. (1997). Adaptive color segmentation – a comparison of neural and statistical methods. *IEEE Transactions on Neural Networks*, Vol. 8, No. 1, (January 1997), pp. 175-185, ISSN 1045-9227

Lowe, D.G. (2004). Distinctive Image Features from Scale-Invariant Keypoints. International Journal of Computer Vision, Vol. 60, No. 2, (November 2004), pp. 91–110, ISSN 1573-1405

Maintz, J.B.A. & Viergever, M.A. (1998). A survey of medical image registration. *Medical Image Analysis*, Vol. 2, No. 1, (March 1998), pp. 1-36, ISSN 1361-8415

Marques de Sá, J.P. (2001). *Pattern Recognition: Concepts, Methods and Applications*, Springer, ISBN 3540422978, Berlin, Germany

Marr, D. & Hildreth, E. (1980). Theory of Edge Detection. *Proceedings of the Royal Society of London. Series B, Biological Sciences*, Vol. 207, No. 1167, (February 1980), pp. 187-217, ISSN 00804649

Neumann, R. & Stanley, C.J. (2008). Specular reflectance data for quartz and some epoxy resins: implications for digital image analysis based on reflected light optical microscopy, *ICAM 2008 - Ninth International Congress for Applied Mineralogy Conference Proceedings*, pp. 703-706, ISBN 9781920806866, Brisbane, Australia, September 8-10, 2008

Orchard, M.T.; Bouman, C.A. (1991). Color Quantization of Images. *IEEE Transactions on Signal Processing*, Vol. 39, No. 12, (December 1991), pp. 2677-2690, ISSN 1053-587X

Otsu, N. (1979). A Threshold Selection Method from Gray-Level Histograms. *IEEE Transactions on Systems, Man, and Cybernetics*, Vol. 9, No. 1, (January 1979), pp. 62-66, ISSN 0018-9472

Paciornik, S. & Gomes, O.D.M. (2009). Co-site Microscopy: Case Studies. *Praktische Metallographie*, Vol. 46, No.9, (September 2009), pp. 483-498, ISSN 0032-678X

Paciornik, S. & Maurício, M.H.P. (2004). Digital Imaging, In *ASM Handbook, Volume 9: Metallography and Microstructures*, G.F. Vander-Voort (Ed.), pp. 368-402, ASM International, ISBN 0871707063, Materials Park, USA

Petruk, W. (1988). The capabilities of the microprobe Kontron image analysis system: application to mineral benefication. *Scanning Microscopy*, Vol. 2, No. 3, pp. 1247-1256, ISSN 0891-7035

Piller, H. (1996). Colour measurements in ore-microscopy. *Mineralium Deposita*, Vol. 1, No. 3, (December 1966), pp. 175-192, ISSN 0026-4598

Pirard, E.; Lebrun, V. & Nivart, J.-F. (1999). Optimal Acquisition of Video Images in Reflected Light Microscopy. *Microscopy and Analysis*, Vol. 60, (July 1999), pp. 9–11, ISSN 0958-1952

Pirard, E. (2004). Multispectral imaging of ore minerals in optical microscopy. *Mineralogical Magazine*, Vol. 68, No. 2, (April 2004), pp. 323–333, ISSN 0026-461X

Pun, T. (1981). Entropic thresholding, a new approach. *Computer Graphics and Image Processing*, Vol. 16, No. 3, (July 1981), pp. 210-239, ISSN 0146-664X

Reimer, L. (1998). *Scanning Electron Microscopy: Physics of Image Formation and Microanalysis* (2nd Edition), Springer-Verlag, ISBN 3540639764, Berlin, Germany.

Schowengerdt, R.A. (1983). *Techniques for image processing and classification in remote sensing*, Academic Press, ISBN 0126289808, Orlando, USA

Serra, J. (1982). Image Analysis and Mathematical Morphology, Academic Press, ISBN 0126372403, London, UK

Serra, J. (1988). Image Analysis and Mathematical Morphology, Volume 2: Theoretical Advances, Academic Press, ISBN 0126372411, London, UK

Sharma, G. & Trussell, H.J. (1997). Digital color imaging. *IEEE Transactions on Image Processing*, Vol. 6, No. 7, (July 1997), pp. 901-932, ISSN 1057-7149

Soto, O.A.J.; Gomes, O.D.M.; Pino, G.A.H. & Paciornik, S. (2004). Native Copper Analysis through Digital Microscopy, In *Applied Mineralogy: Developments in Science and Technology*, M. Pecchio, F.R.D. Andrade, L.Z. D'Agostino, H. Kahn, L.M. Sant'Agostino, M.M.M.L. Tassinari (Eds.), Vol. 2, pp. 1043-1046, ISBN 859865602X

Sutherland, D. & Gottlieb, P. (2001). Application of automated quantitative mineralogy in mineral processing. *Minerals Engineering*, Vol. 4, No. 7-11, pp. 753-762, ISSN 0892-6875

Szeliski, R. (2004). *Image alignment and stitching: a tutorial*, Technical Report MSR-TR-2004-92, Microsoft Research, December 2004.

Theodoridis, S. & Koutroumbas, K. (2003). *Pattern Recognition* (2nd edition), Academic Press, ISBN 0-12-685875-6, London, UK

Toussaint, G.T. (1974). Bibliography on Estimation of Misclassification. *IEEE Transactions on Information Theory*, Vol. 20, No. 4, (July 1974), pp. 472-479, ISSN 0018-9448

Valev, V. & Asaithambi, A. (2001). Multidimensional pattern recognition problems and combining classifiers. *Pattern Recognition Letters*, Vol. 22, No. 12, (October 2001), pp. 1291-1297, ISSN 0167-8655

Vandenbroucke, N.; Macaire, L. & Postaire, J.G. (2003). Color image segmentation by pixel classification in an adapted hybrid color space. Application to soccer image analysis. *Computer Vision and Image Understanding*, Vol. 90, No. 2, (May 2003), pp. 190-216, ISSN 1077-3142

van den Elsen, P.A.; Pol, E.J.D. & Viergever, M.A. (1993). Medical Image Matching – A Review with Classification. *IEEE Engineering in Medicine and Biology*, Vol. 12, No. 1, (March 1993), pp. 26-39, ISSN 0739-5175

Zitova, B. & Flusser, J. (2003). Image registration methods: a survey. *Image and Vision Computing*, Vol. 21, No. 11, (October 2003), pp. 977-1000, ISSN 0262-8856

2

Cutting Mechanism of Sulfurized Free-Machining Steel

Junsuke Fujiwara

Osaka University,

Japan

1. Introduction

In order to improve efficiency of cutting process in production industry, development of new steel which has good machinablity is desired. The work material which has good surface roughness, easy breakable chip and small tool wear as the good machinablity is expected. And the free-machining steel was developed owning to adding elements which could make the machinablity better. Of all others, leaded free-machining steel and sulfurized free-machining steel are famous. The leaded free-machining steel and sulfurized free-machining steel are well used in the production industry. However the use of the leaded free-machining steel is limited from an environmental problem. In order to develop new environmental friendly free-machining steel, it is necessary to find out the behavior of the inclusion in the work material for the improvement of the machining performance.

There are a lot of studies about the behavior of the inclusion in the free-machining steel (Narutaki et al., 1987), (Yaguchi, 1991), (Usui et al., 1980). There are some papers about the role of the lead and the manganese sulfide which are the representative inclusions. The Pb inclusion acts as lubricant and reduces cutting resistance (Akazawa, 1997). As the MnS is harder than steel, the MnS acts as an internal stress concentration source when the work material reforms into a chip at the cutting edge. And the MnS produced the micro-cracks at shear deformation zone. This is the cause that the shear area became small and reduces the cutting stress (Yamamoto, 1971). Although these results are almost reasonable, we must think over the role of the inclusion again in order to produce new free-machining steel. The experiment was carried out to find out the mechanism of the sulfurized inclusion on the machinablity, using some kinds of steels which have different size of the inclusion. The observation of the deformation behavior near the cutting edge was carried out to investigate the effect of the inclusion in detail.

2. Experimental method

In this experiment, two kinds of the sulfurized free-machining steels (Steel A and Steel B) which have different size of the inclusion were used. Figure 1 show optical microphotographs of microstructure and size distribution of MnS in the Steel A and Steel B, respectively. The area fraction of equivalent circle diameter of the inclusions was also shown in these figures. The steel A contains larger inclusions than the steel B. These sulfurized free-machining steels contain 0.42% S, and chemical compositions of these materials are almost the same as shown in Table 1. The inclusions tend to be slender parallel to rolling direction.

In an orthogonal cutting at low speed, the cutting forces were measured. The cutting width of the work material was 2 mm. The surface of work materials was polished to observe deformation of the inclusions. An orthogonal cutting was performed using table feeding system of a horizontal milling machine as shown in Fig. 2. Table 2 shows cutting conditions in the orthogonal cutting. The cutting speed was 16mm/min and the depth of cut was 0.1mm. The tool material was high speed steel and its rake angle was 10 degree.

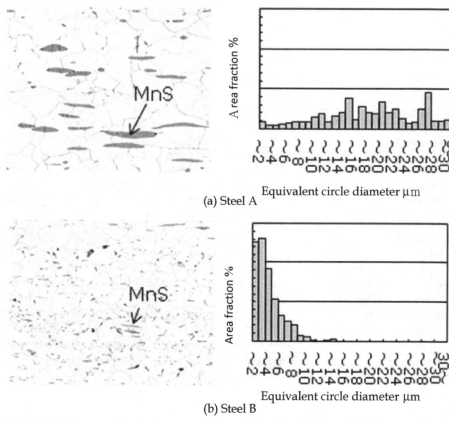

(a) Steel A

(b) Steel B

Fig. 1. Optical micrographs of microstructure and size distribution of MnS

Mass %	C	Si	Mn	S	Al	O2
Steel A	0.03	0.01	1.44	0.42	0.001	0.0175
Steel B	0.03	0.01	1.7	0.43	0.001	0.0044

Table 1. Chemical compositions of work materials

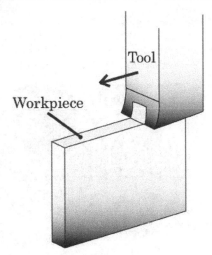

Fig. 2. Method of orthogonal cutting

Cutting speed	0.016 m/min
Depth of cut	0.1 mm
Tool	SHK4
Rake angle	10°
Clearance angle	17°

Table 2. Cutting conditions in orthogonal cutting

3. Experimental results and discussions

3.1 Orthogonal cutting

The Cutting forces were measured in the orthogonal cutting. These results are shown in Fig. 3. On the whole, the cutting force in the Steel A was bigger than that in the Steel B. The cutting force in the Steel A was more stable than that in the Steel B. This fact led smooth surface roughness.

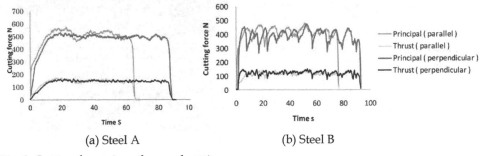

(a) Steel A (b) Steel B

Fig. 3. Cutting forces in orthogonal cutting

In order to investigate the flow state of around the shear zone, a quick stop test was carried out during the orthogonal cutting. Figure 4 shows the enlarged photographs around the shear zone. In case of the Steel A, the large crack parallel to the shear plane was found. In case of Steel B, the chip is thin, and there are small cracks near the rake face.

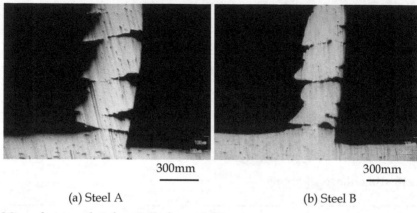

300mm   300mm

(a) Steel A   (b) Steel B

Fig. 4. Microphotographs of partially formed chip

3.2 Micro-cutting in SEM

A small orthogonal cutting equipment as shown in Fig. 5 was mounted into the Scanning Electrical Microscope (SEM) (Iwata 1977). The deformation behavior around the shear zone was observed in detail with the SEM. An example of the micro-cutting in SEM is also shown in this figure. The cutting speed was 0.27 mm/s and the depth of cut was 20 - 50 μm. Table 3 shows the cutting conditions. The work material was cut from the test piece as shown in Fig.6.

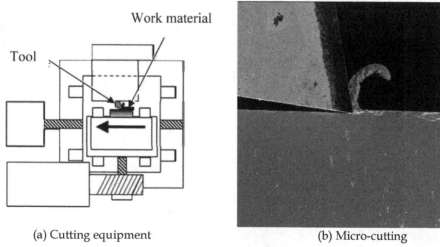

(a) Cutting equipment   (b) Micro-cutting

Fig. 5. Cutting equipment in SEM and micro-cutting

Cutting speed	0.27 mm/s
Depth of cut	20~50 μm
Tool	SKH4
Rake angle	6°
Clearance angle	3°

Table 3. Cutting conditions in SEM

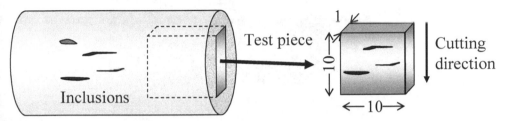

Fig. 6. Work material for micro-cutting in SEM

Figure 7 shows the sequential photographs during cutting of the Steel A. As the Steel A has the spindle shaped inclusions, the inclusion MnS which was extended perpendicular to the cutting direction can be found. This inclusion turned in counterclockwise and broke to several pieces around the shear zone. These pieces create voids around them, and flowed to the chip in the direction parallel to the shear plane. As shown in this figure, the void was formed at the upper of the inclusion and the micro-crack was formed along the primary shear plane.

Figure 8 shows the sequential photographs during cutting of the Steel B. The inclusion of higher aspect ratio than that in the Steel A can be found. As the Steel B had long slender inclusions, this inclusion broke into smaller pieces than that in the Steel A. These pieces create very small voids between them. The inclusions in the Steel B are well dispersed, so these very small voids are created at various places in the work material. It causes the reduction of the cutting force.

3.3 Image analysis for stress distribution

It is very important to know strain and stress distribution in shearing zone. The sequential images could be taken during micro-cutting in SEM. Using with image processing, the strain increase and stress distribution around MnS can be calculated. That is to say, as comparison with two sequential SEM images after micro movement of the tool, the displacement within observed zone could be measured by tracing a same point. Moreover the strain increase and stress distribution could be calculated from the displacement.

In order to measure the displacement from the sequential SEM images, PIV (Particle Image Velocimetry) method was used (Raffel etal., 2007). The moving distance was calculated from gray level pattern between SEM image A at t in time and SEM image B at t+$\triangle$t in time as shown in Fig.9.

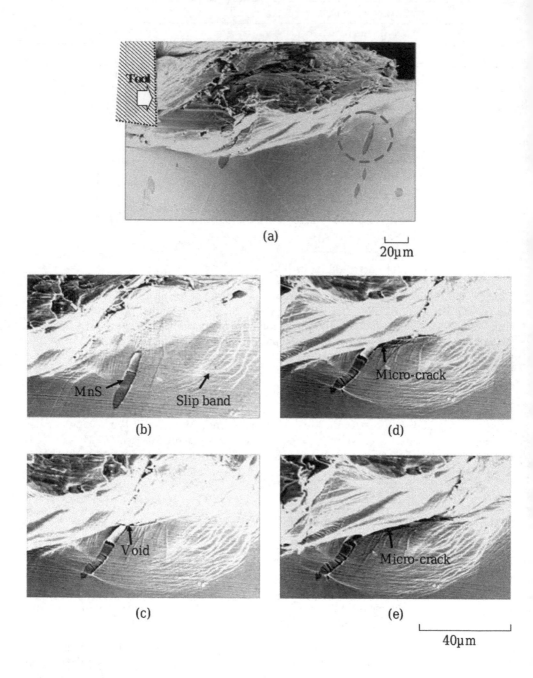

(a)

20μm

(b)

MnS   Slip band

(d)

Micro-crack

(c)

Void

(e)

Micro-crack

40μm

Fig. 7. Deformation behavior of large spindle shaped MnS inclusion (Steel A)

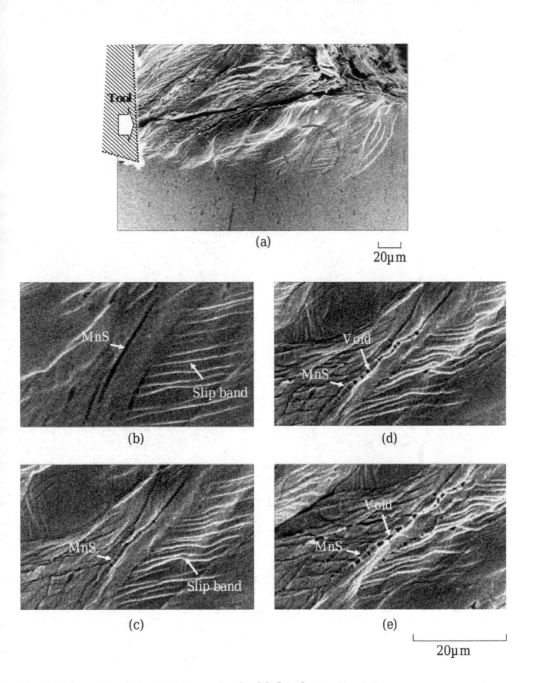

Fig. 8. Deformation behavior of long slender MnS inclusion (Steel B)

The strain and stress distribution was calculated from the displacement increase measured with the PIV method using with FEM as shown in Fig.10 (Usui et al., 1990). In the calculation, the mechanical properties as shown in Table 4 were used.

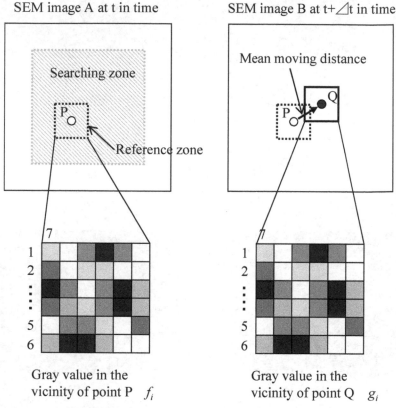

Fig. 9. Outline of correlation method in particle image velocimetry

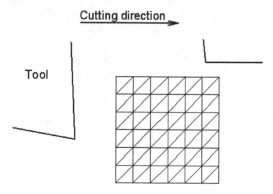

Fig. 10. Model of triangulated zone for FEM calculation

	Steel A	Steel B
Yield strength MPa	522	560
Tensile strength MPa	525	563
Elongation %	13.2	12.0
Reduction of area %	52.1	48.0

Table 4. Mechanical properties of work materials

The strain increase distribution was calculated from the two sequential SEM Images as shown in Fig. 7. Figure 11 shows the strain increase distribution in micro-cutting of the Steel A. The moving distance of two images was 1.8 μm. As shown in Fig 11 (d), the shear strain increase was large in shear zone but another large strain increase was found around MnS.

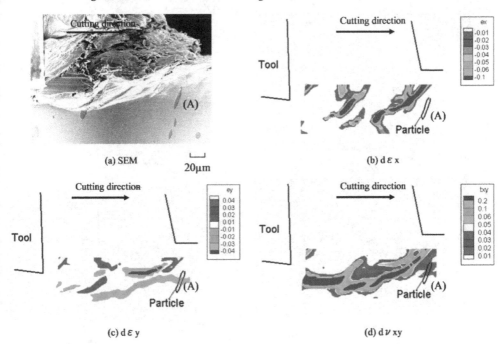

Fig. 11. SEM image and strain increase distribution in micro-cutting of Steel A

The stress distribution was calculated from the strain increase distribution. Figure 12 shows the stress distribution in micro-cutting of the Steel A. It shows that the stress was big at the upper of the MnS because of the stress concentration.

The strain increase distribution was calculated from the two sequential SEM Images as shown in Fig. 8. Figure 13 shows the strain increase distribution in micro-cutting of the Steel B. The moving distance of two images was 4.3 μm. As shown in Fig 13 (d), the strain increase along shear zone was large in and no large strain increase was found around MnS.

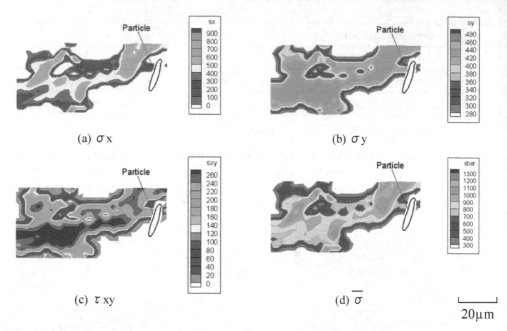

Fig. 12. Stress distribution in micro-cutting of Steel A

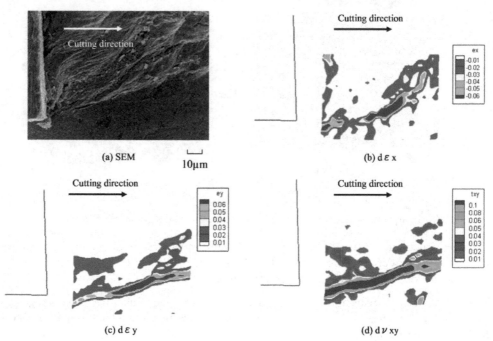

Fig. 13. SEM image and strain increase distribution in micro-cutting of Steel A

The stress distribution was calculated from the strain increase distribution. Figure 14 shows the stress distribution in micro-cutting of the Steel B. As the Steel B had long slender inclusions, the inclusion MnS broke into small pieces and the stress distributed along the shear plane. There is little stress concentration around the MnS.

Consequently, in the Steel A which has large spindle type MnS, the micro-crack is easily formed. As this micro-crack affects the breakage of a build-up edge (BUE) and chip, the BUE could not become big and the good finish surface roughness was gained in the Steel A. As shown in Fig. 15, large micro-crack affected the separation between the chip and the BUE.

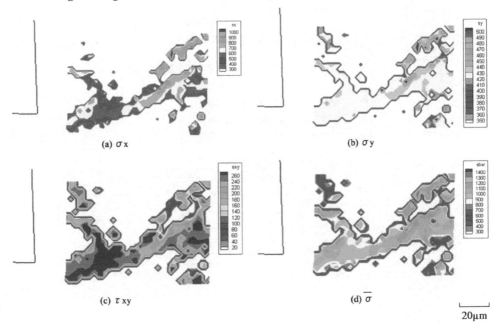

(a) σx

(b) σy

(c) τxy

(d) $\bar{\sigma}$

20μm

Fig. 14. Stress distribution in micro-cutting of Steel A

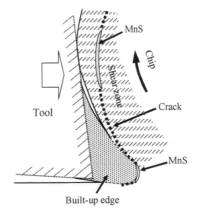

Fig. 15. Effect of spindle-type MnS to suppress BUE growth in the vicinity of BUE

3.4 Quick stop test during turning.

A quick stop experiment during turning was carried out with the quick stop system which was attached to a conventional lathe. Figure 16 shows the equipment of the quick stop test. The tool was rotated down at the moment when pulling a pin which fixed the tool and the cutting state was stopped quickly. The deformation behavior around the shear zone was observed from the workpiece with a chip. The cutting speed was 62 m/min and the depth of cut was 0.2 mm. Table 5 shows the cutting conditions.

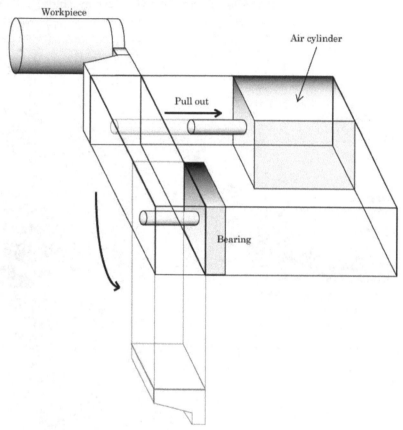

Fig. 16. Equipment of quick stop test

Cutting speed	62 m/min
Depth of cut	0.2 mm
Tool	Cemented carbide P10
Rake angle	20°
Clearance angle	6°

Table 5. Cutting condition in quick stop test

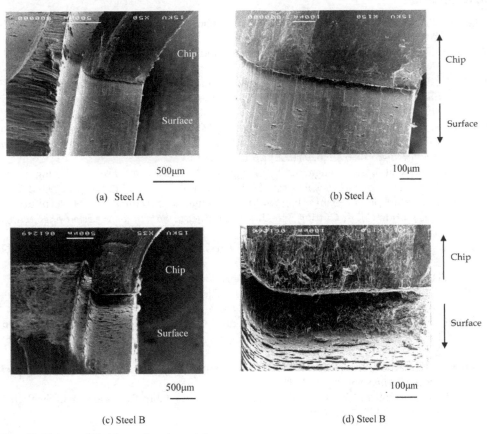

(a) Steel A   (b) Steel A

(c) Steel B   (d) Steel B

Fig. 17. Chip and finish surface in quick stop test

In quick stop test, the chips in the vicinity of the tool face and machined surface was observed as shown in Fig. 17. In case of the Steel A which contains large spindle shaped inclusions, the BUE could not be found. And machined surface had good surface roughness. In case of the Steel B which contains small slender inclusions, the BUE could be found on the rake face of the chip. There were many tears on the machined surface. The BUE partially separated and they leave on the machined surface. As a result, the surface roughness became bad. In this experiment, it was clear that the larger inclusions could reduce the formation of the BUE.

4. Conclusions

The main results obtained are as follows.

1. In machining of sulfurized free-machining, some inclusions creates voids around them, some break to several pieces depending on their conditions around the shear zone.
2. The larger inclusions can reduce the formation of the BUE.

5. References

Akazawa, T., Free cutting steels contributing to industry, *Journal of Special Steel*, Vol.46, No.5 (May 1997), pp.6-10.

Araki, T., Yamamoto, S., Machinability of Steel and Metallugical Factors, *Iron and Steel*, Vol.57, No.13 (Nov. 1971), pp.1912-1932.

Iwata, K., Ueda, K., Shibasaka, T, Study on Micro-machining Mechanics Based on Direct-SEM Observation, *Journal of Japan Society of Precision Engineering*, Vol. 43, No.3 (1977) pp.311-317.

Katayama, S., Toda, M., Hashimura, M., Growing Model of Build-up Edge in Relation to Inhomogeneities of Steel Microstructure, *Journal of Japan Society of Precision Engineering*, Vol. 62, No. 9 (1996) pp.1345-1349.

Maekawa, K., Kubo, A., Kitagawa, T., MachinabilityAnalysis of Free-machining Steel, *Journal of Japan Society of Precision Engineering*, Vol. 57, No. 12 (1991) pp.2193-2198.

Narutaki, N., Yamane, Y., Usuki, H., Yan, B., Kuwana, T., Machinability of Resulfurized Steels under High Cutting Speed, *Journal of Japan Society for Precision Engineering*, Vol.53, No.3 (March 1987), pp.455-466.

Raffel, M., Willert, C.E., Kompenhaus J., (2000) Particle Image Velocimetry, Springer, ISBN 3-540-63683-8 New York.

Usui, E., (1990) *Modern Cutting Theory* (1st), Kyouritsu Shuppan, ISBN 4-320-08054-8, Tokyo.

Usui, E., Obikawa, T., Shirakashi, T., Embrittle Action of Free-machining Additives, *Japan Society for Precision Engineering*, Vol.46, No.7 (July 1980), pp.849-855.

Yaguchi, H., Effect of MnS Inclusion Size on Machinability of Low-Carbon Leaded Resulfurized Free-Maching Steel, *Journal of Applied Metalworking* Vol. 4, No. 3 (1986) pp.214-220.

Yaguchi, H., Effect of soft assitives (Pb, Bi) on formation of Build-up edge, *Journal of Material Science Technology*, No. 4, (1988) pp.926-932.

Yaguchi, H., The Role of Liquid Metal Embrittlement on the Chip Disposability of Steel, *Iron and Steel* Vol.77, No.5 (May 1991), pp.683-690.

3

SEM Analysis of Precipitation Process in Alloys

Maribel L. Saucedo-Muñoz,
Victor M. Lopez-Hirata and Hector J. Dorantes-Rosales
Instituto Politecnico Nacional (ESIQIE),
Mexico

1. Introduction

The microstructural characterization of the precipitation process in alloys is a very important aspect in order to understand the formation mechanism and growth kinetics of precipitated phases during its heating because of either the heat treating process or the operation-in-service conditions. Additionally, the microstructure control is a key point to know the degree of hardening after heat treating of the alloys and to assess their mechanical properties after a prolonged exposure at high temperature during the operation of an industrial component. There are different characterization techniques for microstructure; however, the use of the scanning electron microscopy, SEM, has been very popular for the microstructural observation and it has become a power tool for characterization of the phase transformations. Besides, the application of energy-dispersed-spectra, EDS-SEM system to the microstructural characterization has permitted to know not only the morphology of phases, sizes, distribution and then growth kinetics, but also their chemical composition and thus element distribution of the formed phases. Thus this chapter shows the application of SEM-EDS system to the characterization of microstructural of precipitation process in different alloy systems such as Fe-Ni-Al alloy, austenitic stainless steels and Mg-Zn-Al alloy.

2. Precipitation in alloys

Phase separation in alloys usually consists of the formation of a supersaturated solid solution by heating the alloy at temperatures higher than the equilibrium solvus line and subsequently quenched rapidly. This supersaturated solid solution can usually be separated in two or more phases as a result of the isothermal aging at temperatures lower than that of equilibrium. Phase separation can mainly take place by two mechanisms, nucleation and growth, and spinodal decomposition (Porter, 2009). The former mechanism consists of the formation of a stable nucleus with a nucleation barrier to overcome and it is characterized by an incubation period. In contrast, the latter one is initiated by the spontaneous formation and subsequent growth of coherent composition fluctuations. The formation of fine second-phase dispersion in a matrix promotes its hardening, known as precipitation hardening. If the aging of alloys continues, it is expected that larger precipitates will grow at the expense of smaller ones which dissolve again given rise to a change in the precipitate size distribution (Kostorz, 2005).

2.1 Coarsening process in Fe-Ni-Al alloys

The precipitation of the β´phase is important for strengthening at high temperatures in different engineering ferritic alloys such as, PH stainless steels, nitralloy, Fe-Cr-Ni-Al based alloys, etc. These alloys are used in industrial components which require good mechanical strength and oxidation resistance at high temperatures. The β´phase is an ordered phase of the B2 type crystalline structure (Sauthoff, 2004). The coarsening resistance of precipitates is a key factor to keep the high strength at high temperatures in this type of alloys. An alternative to have a good coarsening resistance, it is to have a low value of lattice misfit which maintains a coherent interface between the precipitate and matrix (Kostorz, 2005). Thus, this section shows the effect of structural and morphological characteristics of the β´ precipitates on the coarsening behavior during the isothermal aging of an Fe-10Ni-15Al alloy.

2.1.1 Experimental details

An Fe-10Ni-15Al alloy (wt. %) was melted using pure metallic elements in an electrical furnace under an argon atmosphere. The ingot of 30 x 10 x 10 mm was encapsulated in a quartz tube with argon gas and then homogenized at 1100 °C for one week. Specimens were solution treated at 1100 °C for 1 h and subsequently aged at temperatures of 750, 850 and 920 °C for times from 0.25 to 750 h. These samples were also observed with a SEM analysis with EDS detector at 20 kV. Vickers hardness was tested for the aged specimens using a load of 100 g.

2.1.2 Microstructural evolution of coarsening

SEM micrographs of precipitates are shown for the sample aged at 750 and 920 °C for different times in Figs. 1 (a-c) and (d-f), respectively. The shape of the β´ precipitates was round without any preferential alignment for the aging at 750 °C up to 75 h and 920 °C up to 0.5 h, Figs 1 (a). A further aging changed the shape of the β´ precipitates to cuboids with a preferential alignment on the <100> directions of the ferritic α phase, Figs. 1 (c-e). A prolonged aging at 920 °C promoted the change of shape to rectangular plates also aligned in the <100> directions, Fig. 1 (f). The volume percentage of precipitation was determined to be about 30, 25 and 20 % for the samples aged at 750, 850 and 920 °C, respectively.

2.1.3 Growth kinetics of coarsening

The variation of the β´ precipitates size expressed as $r^3-r_o^3$ with aging time for the sample aged at 750, 850 and 920 °C is shown in Fig. 2. It can be noticed that the experimental data fit to a straight line for each temperature. Thus the growth kinetics of coarsening followed the behavior predicted by the Lifshitz-Slyozov-Wagner (LSW) theory for coarsening controlled by volume diffusion. This fact shows a good agreement with the modified theory for the diffusion-controlled coarsening in ternary alloys (Kostorz, 2005) which predicts that growth kinetics is similar to that of LSW theory. The size distribution of precipitates is shown in Figs. 3 (a-c) for the sample aged at 750, 850 and 920 °C for 200 h, respectively. It can be seen that the size distribution is broader and lower than that predicted by the LSW theory because of the high volume fraction of precipitates, which has been reported in the coarsening process of several alloy systems. It has been observed that the growth or shrinkage rate of an individual particle depends not only on its normalized radii but also on its local environment. That is, a particle surrounded by several larger particles will grow slower, or shrink faster, than a particle of the same size whose neighbors are smaller. Thus, as the volume fraction increased, the particle size distribution widened increasing the coarsening rate at the same time. It was also observed that the higher aging temperature, the faster coarsening kinetics of the β´ precipitates because of the increase in volume diffusion (Ratke & Vorhees, 2002).

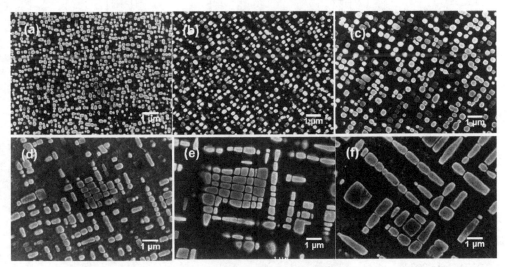

Fig. 1. SEM micrographs for the Fe-10Ni-15Al alloy aged at 750 °C for (a) 75, (b) 250, and (c) 500 h, and at 920 °C for (d) 25, (e) 100 and (f) 200 h.

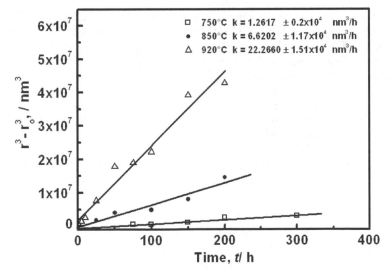

Fig. 2. Plot of r^3-r_o^3 vs. aging time for the Fe-10Ni-15Al alloy aged at 750, 850 and 920 °C.

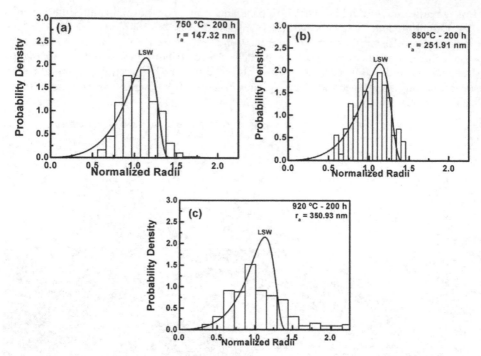

Fig. 3. Size distribution of precipitates for the Fe-10Ni-15Al alloy aged at (a) 750, (b) 850 and (c) 920 °C for 200 h.

2.1.4 Hardening behavior

Figure 4 shows the aging curves for the sample aged at 750 and 920 °C. A higher hardness can be noticed in the sample aged at 920 °C. This can be attributed to the morphology and alignment of β′ precipitates. That is, they are cuboids aligned in the <100> crystallographic directions of the ferritic matrix. A similar hardening behavior was observed in Fe-Ni-Al alloys aged at lower temperatures, 500 °C (Soriano-Vargas et al. 2010, Cayetano-Castro et al. 2008). In contrast, the precipitates are rounded particles without any preferential crystallographic alignment for aging at 750 °C up to 75 h. Besides, the size of β′ precipitates is much smaller than that of the sample aged at 920 °C. It can also be observed that the hardness peak is first reached in the aging at 750 °C than that at 920 °C. Additionally, the overaging started first for the aging at 750 °C. Besides, the hardness is almost the same value for prolonged aging at both temperatures. All the above facts suggest that even the coarsening process at 920 °C is the fastest one, the cuboid morphology and alignment of β′ precipitates causes a higher hardness peak and a slower overaging process than those corresponding at 750 °C.

In summary, the aging process of the Fe-10Ni-15Al alloy promoted the precipitation of the β′(Fe(NiAl)) precipitates with the B2 type crystalline structure. The morphology of β′ precipitates was rounded at the early stages of aging and then it changed to cuboids aligned in the <100> directions of the ferritic matrix. A prolonged aging caused the formation of rectangular plates also aligned in this direction. The coarsening process followed the growth kinetics predicted by the LSW theory. Nevertheless, the hardness peak was higher and the overaging process occurred later in the sample aged at 920 °C than those of the sample aged at 750 °C. This behavior can be attributed to the fast formation of cuboid morphology and alignment in the <100> direction due to the higher lattice misfit between the ferritic matrix and β´ precipitate at this aging temperature.

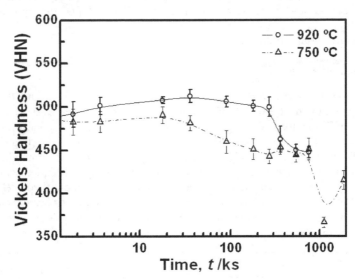

Fig. 4. Aging curves for the Fe-10Ni-15Al alloy aged at 750 and 920 °C.

2.2 Precipitation in austenitic stainless steels

The austenitic stainless steels are construction materials for key corrosion-resistant equipment in most of the major industries, particularly in the chemical, petroleum, and nuclear power industries (Marshal, 1984). These steels are iron alloys containing a minimum of approximately 12 % chromium. This content of chromium allows the formation of the passive film, which is self-healing in a wide variety of environments. Nitrogen as an alloying element in iron-based alloys is known since the beginning of the last century having been profoundly studied during the last three decades (Nakajima et al., 1996). Nevertheless, nitrogen steels are now not widely used. The reason for the comparatively narrow industrial application lies in the old customer skepticism in relation to nitrogen as an element causing brittleness in ferritic steels, some technical problems involved with nitrogen into steel, and the insufficient knowledge of the physical nature of nitrogen in iron and its alloys. In the case of austenitic stainless steels, the main driving force in the development of nitrogen-containing steels is due to the higher yield and tensile strengths achieved, compared with conventionally-processed austenitic stainless steels without sacrificing toughness. Nitrogen stainless steels have yield and tensile strengths as much as 200-350 % of the AISI 300 series steels. It is also important to notice that, in contrast to carbon, nitrogen-containing austenitic stainless steels retain high fracture toughness at low temperatures. Therefore, the higher mechanical properties of nitrogen-containing austenitic stainless steels have made very attractive its application in the power-generation industry, shipbuilding, railways, cryogenic process, chemical equipment, pressure vessels and nuclear industries (Nakajima et al. 1989). These stainless steels are also susceptible to the precipitation of different phases because of the aging for long exposition at high temperatures or during continuous cooling after a welding process. Therefore, it is important to evaluate the degree of microstructural degradation due to the precipitation phenomenon which may affect the cryogenic toughness in this type of steels. In this section, three types of austenitic stainless steels, JJ1, JN1 and JK2 developed for applications to the superconducting magnets of fusion experimental reactor by JAERI, were selected to study the microstructure evolution during isothermal aging.

2.2.1 Experimental details

Materials used in this work were forged-steel plates of 200 mm thick and their chemical compositions are shown in Table 1. The solution treatment of JN1, and JJ1 and JK2 was carried out at 1075 and 1050 °C, respectively, for 1 hour under an argon atmosphere, and then water-quenched. The aging temperatures and times were 600, 700, 800 and 900 °C and from 10 to 1000 minutes, respectively. The aged samples were prepared metallographically and etched with Vilella´s reagent. The precipitates in the aged samples were extracted electrolitically by dissolution of the austenitic matrix in a solution of 10 vol. %HCl-CH$_3$OH at 4 volts. The X-ray diffraction pattern of extracted precipitates was measured in a diffractometer using Kα Cu radiation. The SEM/EDX microanalysis of precipitates was also conducted using the extraction replica technique.

Material	C	Si	Mn	Ni	Cr	Al	N	Mo
JN1	0.040	0.97	3.88	15.07	24.32	0.023	0.32	---
JJ1	0.025	0.48	10.13	11.79	12.01	---	0.236	4.94
JK2	0.05	0.39	21.27	9.15	12.97	---	0.247	0.97

Table 1. Chemical composition (wt.%) of materials.

2.2.2 Microstructural evolution

An intergranular precipitation can be observed for all cases. The highest and lowest volume fraction of intergranular precipitates corresponded to the aged JN1 and JK2 steels, respectively, Figs. 5 (a-b) and (e-f). The presence of an intergranular cellular precipitation of Cr$_2$N was observed to occur in the JN1 steel sample aged at 900 °C. No intergranular precipitation was practically detected for the JK2 steel aged at 700 °C. The intragranular precipitates can be classified into two types: cellular or discontinuous precipitation and plate-like precipitates, which have a preferred alignment with the austenitic matrix. The morphology of cellular precipitates is similar to that of pearlite in carbon-steels, Fig. 6. The formation of this lamellar microstructure initiated at grain boundaries and grew into the austenite γ matrix, according to the following reaction:

$$\gamma \rightarrow \gamma + Cr_2N \qquad (1)$$

The volume fraction of the discontinuous precipitation increased with time and the maximum value was determined by the point-count grid method, to be about 0.04. This value seems to be reasonable, since a volume fraction of 0.1 was reported in an austenitic stainless steel containing 0.42 % N, after a long heat treatment (Kikuchi et al., 1991). Some small intragranular precipitates were present in the JN1 and JJ1 steels aged at 700 and 800 °C for 5 h, Figs. 5 (a-d). The volume fraction of intragranular precipitates for the aged JJ1 steel was slightly higher than that of the aged JN1 steel. This tendency became higher by increasing the aging temperature. Almost no intragranular precipitation was observed in the aged JK2 steel. The precipitation of particles was also observed to occur on twin boundaries for the aged JN1 steel. The X-ray diffraction patterns of residues extracted from the JN1, JJ1 and JK2 steels aged at 700 and 800 °C for 5 h are shown in Fig. 7. The extracted precipitates of the JN1 steel, aged at 700 and 800 °C for 5 h, were identified as $Cr_{23}C_6$ and Cr_2N. The Cr_2N and $Cr_{23}C_6$ phases were also detected in the aged JJ1 steel. Besides, the presence of the (Fe_2Mo) η phase was also noted in the samples aged at 800 and 900 °C. The precipitated particles of JK2 steel were mainly composed of $Cr_{23}C_6$. According to the chemical composition, shown in Table 1, the JN1 steel has the highest and lowest contents of interstitial solutes (C and N), and Mn, respectively. This suggests that the highest volume fraction of precipitation for carbides and nitrides must have occurred in this steel. In contrast, the JK2 steel has an interstitial solute content lower than that of the JN1 steel, but it has the highest content of manganese, which maintains nitrogen in solid solution, avoiding its precipitation. That is, it is only expected the precipitation of carbides for this steel. This fact showed a good agreement with the above results.

2.2.3 Precipitation kinetics

All the above results are summarized in the Time-Temperature-Precipitation (TTP) diagrams of JN1, JJ1 and JK2 steels, as shown in Figs. 8 (a-c), respectively. In general, it can be noticed that the kinetics of precipitation for JN1 steel is faster than that of JJ1 steel, because of its higher interstitial solute content. The TTP diagrams show that the intergranular precipitation of $Cr_{23}C_6$ and Cr_2N preceded to the intragranular precipitation of Cr_2N, and Cr_2N and η phase in JN1 and JJ1 steels, respectively.

2.2.4 Fracture toughness

In contrast, Figs. 9 (a), (b) and (c) show the plots of CVN fracture energy at -196 °C versus aging time for the JN1, JJ1 and JK2 steels aged at 700, 800 and 900 °C, respectively. All the steels showed a monotonotic decrease in the CVN fracture energy with aging time at the three temperatures. It is also evident that the drop of fracture toughness of JN1 steel is always faster than that of JJ1 steel. The fastest drop of fracture toughness occurred in the JN1 steel samples aged at 900 °C. This fact may be attributed to the higher content of C and N in JN1 steel, which can lead to faster kinetics in intergranular precipitation during the aging process, as discussed in a later section. The CVN fracture energy of solution treated JK2 steel was lower than that corresponding to the other two steels. The lowest decrease in the CVN fracture energy was for the aged JK2 steel. Furthermore, the JK2 steel, aged at 900 °C, showed almost no change in the fracture energy with time. All the JN1, JJ1 and JK2 steels fractured in a ductile manner in the solution treated condition. Intergranular facets were found in all the aged samples, although the area fraction of intergranular facets to ductile surface was strongly dependent on aging conditions. The fraction of intergranular brittle fracture increased with time and temperature, and it seemed consistent with the CVN fracture energy value. Nevertheless, the fracture surface of the JK2 steel, aged at 900 °C, showed almost a complete ductile- fracture mode. These results are in agreement with the fracture mode observed in the tested SP test specimens.

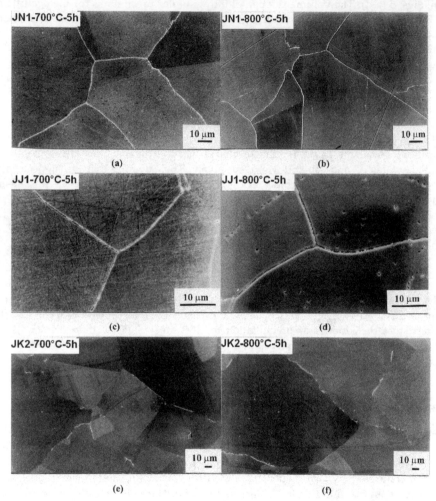

Fig. 5. SEM micrographs of JN1, JJ1 and JK2 steels aged at 700 and 800 °C for 5 h.

In summary, the highest and lowest degradation in toughness for JN1 and JK2 steel, respectively, is associated with the volume fraction of intergranular precipitation formed during the thermal aging. An abundant presence of intergranular precipitates was reported to causes the reduction of cohesive strength of grain boundaries (Saucedo et al., 2001). This is also confirmed by the increase in intergranular brittle fracture as the thermal aging progresses.

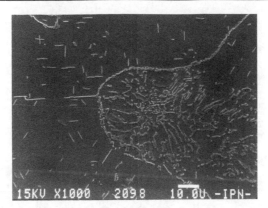

Fig. 6. SEM micrograph of the cellular precipitation in the JN1 steel aged at 700 °C for 1000 h.

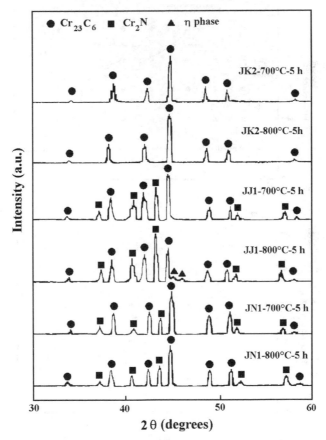

Fig. 7. X-ray diffraction patterns of extracted residues for JN1, JJ1 and JK2 steels agedat 700 and 800 °C for 5 h.

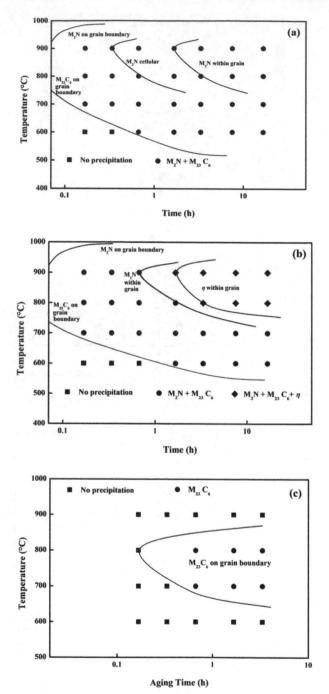

Fig. 8. TTP diagrams of the (a) JN1, (b) JJ1 and (c) JK2 steels.

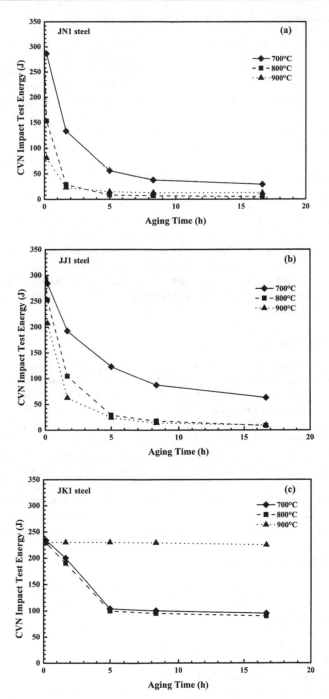

Fig. 9. Plot of CVN impact energy at – 196 °C vs. aging time for tested steels.

2.3 Cellular precipitation in a Mg-8.5Al-0.5Zn-0.2Mn alloy

Mg-Al-Zn alloys have become one of the most important light alloys with a wide range of applications in the automotive industry. This is attributed to the best combination of castability, mechanical strength and ductility (Kainer, 2003). The AZ series of magnesium alloys are mainly based on the Mg-Al binary alloys system. According to the equilibrium Mg-Al alloy phase diagram, the equilibrium phases are the hcp Mg-rich α phase and $Mg_{17}Al_{12}$-γ phase with a complex bcc structure. During the aging process of the Mg-Al based alloys, two types of precipitation are present. That is, discontinuous precipitation takes place on grain boundaries. One of these, intergranular precipitations occurs forming a lamellar structure and it is also known as cellular precipitation. Additionally, continuous precipitation takes place in an intragranular manner and it exhibits more complicated morphologies and orientation relationships than the cellular precipitation. It has been shown in several works (Lai et al., 2008) that these alloys have a poor response to precipitation hardening, compared with precipitation-hardenable Al alloys. Furthermore, the aging hardness is strongly influenced by the morphology, the size and the distribution density of $Mg_{17}Al_{12}$ precipitates. Besides, it has been reported that both discontinuous and continuous precipitations have an effect on the hardness of these alloys (Contreras-Piedras, et al., 2010). Thus, this section shows the mechanism and growth kinetics of cellular precipitation in a Mg-8.5Al-0.5Zn-0.2Mn (wt.%) alloy aged isothermally at 100, 200 and 300 °C for different time periods.

2.3.1 Experimental details

A Mg-8.5Al-0.5Zn-0.2Mn (wt.%) alloy was melted using pure metallic elements under a protective argon atmosphere. Table 1 shows the chemical analysis corresponding to this alloy. Specimens of 10 mm x 10 mm x 10 mm were cut from the ingot and encapsulated in a Pyrex tube under an argon atmosphere. These were homogenized at 430 °C for 3 days and subsequently water-quenched. Homogenized and solution-treated specimens were aged at 100, 200 and 300 °C for different times. The heat-treated specimens were analyzed by X-ray diffraction with copper Kα radiation. These specimens were prepared metallographically and etched with an etchant composed of 19 ml distilled water, 60 ml ethylene glycol, 20 ml glacial acetic acid and 1 ml nitric acid. Etched specimens were observed at 25 kV with a scanning electron microscope equipped with EDS analysis. Vickers hardness was measured in all the heat-treated samples with a load of 100 g. The volume fraction of the discontinuous precipitation was determined from SEM images using a commercial image analyzer.

2.3.2 Microstructural characterization

The X-ray diffraction patterns of the specimens in the conditions of solution-treated and aged at 300 °C for 150 h are shown in Fig. 10. A single-phase is confirmed in the solution-treated specimen, while the appearance of XRD peaks corresponding to the $Mg_{17}Al_{12}$-γ phase are evident in the XRD pattern of the specimen aged at 300 °C for 150 h. No other phases were detected. The presence of these phases for each case is in agreement with the equilibrium Mg-Al phase diagram. Figures 11 (a-i) show the SEM micrographs for the specimens aged at 100, 200 and 300 °C for different time periods. There is a clear competition between the discontinuous and continuous precipitation from the early to the late stages of aging (see, for instance Fig. 11 (e)). Some intragranular precipitates are also observed in these micrographs, Fig. 11 (h). In general, there is a precipitate coarsening as the aging process progresses, Figs. 11 (e-f) and (h-i). The morphology of cellular precipitation at 100 and 200 °C mainly consisted of an S-shape and double-seam morphologies. In contrast, the shape corresponding to 300 °C was mainly a single-seam. It has been reported (Aaronson et al., 2010) that the first morphology occurs at a low temperature ($T<T_m/2$) and it is associated with the free-boundary mechanism and the second one takes place at lower temperature and it is related to the precipitate-assisted mechanism.

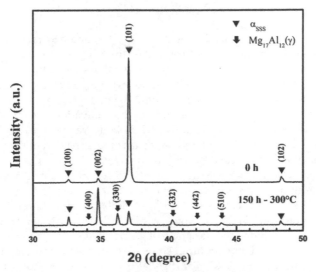

Fig. 10. XRD patterns of the specimens solution-treated and aged at 300 °C for 150 h.

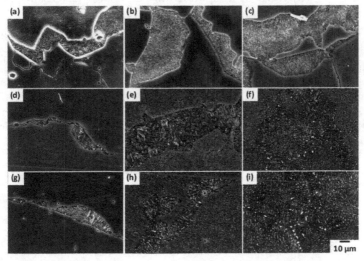

Fig. 11. SEM micrographs of the alloy aged at 100°C for (a) 550, (b) 1500 and (c) 3000 h, At 200 °C for (d) 1, (e) 10 and (f) 250 h, and at 300°C for (g) 0.9, (h) 1 and (i) 25 h.

2.3.3 Growth kinetics of cellular precipitation

The plot of volume fraction of cellular precipitation vs. aging time is shown in Fig. 12. The highest volume fraction occurred for the lowest aging temperature, 100 °C. This fact suggests that continuous precipitation extends more rapidly within grains limiting the growth of cellular precipitation. The analysis of the plot of the volume fraction X_f vs. aging time t, Fig. 12, was carried out using the Johnson-Mehl-Avrami-Kolmogorov equation (Cahn, 1975):

$$X_f = 1 - \exp\left(-kt^n\right) \tag{2}$$

The time exponent n was determined to be about 1.1, 0.85 and 0.87 for 100, 200 and 300 °C, respectively. These exponent values are close to 1 and it is associated with the dimensionality of the saturation site. That is, it corresponds to a boundary (Cahn, 1975). The lamellar structure always nucleates at grain boundaries and grows perpendicularly to them. The cellular growth stops only if the volume fraction of continuous precipitation is significant to impede its growth. In addition, the activation energy for the cellular precipitation was determined to be about 64.6 kJ mol[-1]. It was obtained by the slope of the straight line in the Arrhenius plot of the time for a volume fraction of 0.6 vs. the reciprocal value of the absolute temperature as shown in Fig. 5. This energy value seems to be reasonable because it is much lower than the self-diffusion of Mg, 135 kJ mol[-1] (Mehrer, 1990). That is, it seems to correspond to a grain boundary diffusion process. Additionally, an energy value of 84 kJ mol[-1] was reported for the cellular precipitation in the binary Al-Zn alloy system (Contreras et al., 2010), which is also a low energy value as that found in present work. Figure 6 shows the variation of interlamellar spacing, S, of discontinuous precipitation as a function of temperature. An increase in lamellar spacing is observed with the increase in temperature. A similar behavior was reported for the discontinuous precipitation in Al-Zn alloys (Contreas et al. 2010). According to the Turnbull theory for cell growth kinetics, the interlamellar spacing S is defined as follows (Aaronson et al. 2010):

$$S = -4\gamma V/\Delta G \tag{3}$$

Where γ is the interfacial energy, V the molar volume, and ΔG the free energy associated with the cellular reaction. ΔG has an inverse relation with undercooling, temperature. Thus, the lower temperature corresponds to the shorter interlamellar spacing. Moreover, the interlamellar spacing remains constant with the increase in aging time for all aging temperatures. These facts seem to be in agreement with the Turnbull theory, which predicts constant lamellar spacing and lamellae growth rate according to the following equation (Aaronson et al., 2010):

$$G \approx {\sim}4\delta D_\delta/S^2 \tag{4}$$

Where G is the lamellae growth rate, D_b is the solute diffusivity along the cell boundary and δ is the cell boundary thickness.

2.3.4 Hardenin behavior

The aging curves for 100, 200 and 300 °C are shown in Fig. 7. The lowest and fastest hardness peak was observed in the aging at 300 °C. This behavior can be attributed to the rapid coarsening of the $Mg_{17}Al_{12}$-γ precipitates either in the discontinuous or continuous precipitations. In contrast, the highest and slowest hardness peaks occurred in the alloy aged at 100 °C. This fact seems to be related to the fine continuous precipitation due to the slow diffusion process at this temperature.

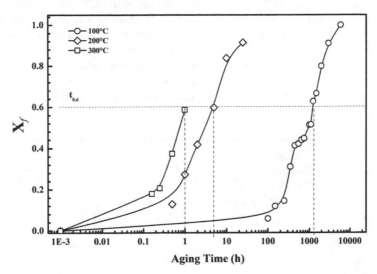

Fig. 12. Volume fraction of cellular precipitation vs. aging time of the alloy aged at 100, 200 and 300 °C.

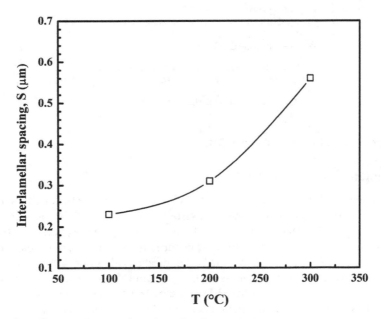

Fig. 13. Interlamellar spacing as a function of aging temperature.

In summary, the microstructural evolution and growth kinetics were studied in an isothermally-aged Mg-8.5Al-0.5Zn-0.2Mn (wt%) alloy and the growth kinetics of cellular precipitation was evaluated using the Johnson-Mehl-Avrami-Kolmogorov equation analysis (Cahn, 1975), which gives a time exponent close to 1. This value confirms that cellular precipitation takes place on the saturation sites corresponding to grain boundaries. Additionally, the activation energy for the cellular precipitation was determined to be about 64.6 kJ mol[-1]. This also indicates a grain boundary diffusion process. The variation of cellular spacing with temperature follows the behavior expected by Turnbull theory. The highest hardness peak corresponded to the lowest aging temperature and it is associated with a fine continuous precipitation, while the lowest hardness peak was detected at the highest aging temperature and it is attributed to the rapid coarsening process of both precipitations.

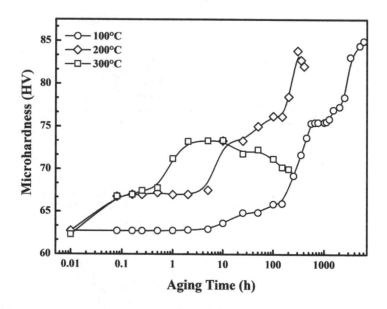

Fig. 14. Aging curves for 100, 200 and 300 °C.

3. Conclusion

This chapter showed three applications of SEM characterization for the analysis of different phase transformations in ferrous and nonferrous alloys, as well as its effect on their mechanical properties. The analysis of these phase transformations enables us to characterize the growth kinetics of these transformations which can be useful either to design heat tretments in order to obtain better mechanical properties or to analyze the microstructural evolution in order to asses the mechanical properties of a component-in-service. Besides, it was shown that the SEM characterization parameters can be used along with the phase transformation theories permitting a better understanding of the transformation behavior in materials after heating.

4. Acknowledgment

The authors wish to acknowledge the financial support from Instituto Politecnico Nacional (ESIQIE), SIP-IPN and CONACYT 100584.

5. References

Aaronson, H.I, Enomoto M. & Lee, J.K. (2010). *Mechanism of Diffusional Phase Transformations in Metals and Alloys*, CRC Press, ISBN 978-1-4200-6299-1, NW, USA

Cayetano-Castro, N.; Dorantes-Rosales H., Lopez-Hirata, V.M., Cruz-Rivera, J. & Gonzalez-Velazquez, J.L. (2008). Cinética de Engrosamiento de Precipitados Coherentes en la Aleación Fe-10%Ni-15%Al. *Revista de Metalurgia de Madrid*, Vol. 44, No. X, (Month, 2008) pp. 162-169, ISSN 1582-2214

Christian J.W. (1975), *The Theory of Transformations in Metals and Alloys*, Pergamon Press, ISBN 0-08-018031-0, Oxford, UK

Contreras-Piedra, E., Esquivel-Gonzalez, R., Lopez-Hirata, V.M., Saucedo-Muñoz, M.L., Paniagua-Mercado, A.M. & Dorantes-Rosales, H.J. (2010). Growth Kinetics of Cellular Precipitation in a Mg-8.5Al-0.5Zn-0.2Mn (wt.%) Alloy, *Materials Science Engineering A*, Vol. 527, pp. 7775-7778, 2010. ISSN 0921-5093

Kainer, K.U. (2003), Magnesium- Alloys and Technologies, Wiley-VCH, ISBN 3-527-30570-X, Germany

Kikuchi M., Kajihara M. & Choi S. (1991). Cellular Precipitation Involving both Substitutional and Interstitial Solutes: Cellular of Cr_2N in Cr-Ni Austenitic Steels. *Materials Science Engineering A*, Vol. 146, pp. 131-150, ISSN 0921-5093

Kostorz, G. (2001). *Phase Transformations in Materials*, Wiley-VCH, ISBN 3-527-30256-5, Weinheim, Germany

Lai W.J.; Lai, Y.Y. Lu, Y.F. Hsu, S. Trong, W.H. Wang. (2009). Aging behaviour and precipitate morphologies in Mg–7.7Al–0.5Zn–0.3Mn (wt.%) alloy, *Journal of Alloys Compounds*, Vol. 476, pp.118-124, ISSN 0925-8388

Marshal, P. (1984). *Austenitic Stainless Steels Microstructure and Properties*, Elsevier Applied Science Publisher, ISBN 0267-0836, NY, USA

Mehrer, H. (1990), *Numerical Data and Functional Relationship in Science and Technology*, Landolt-Borstein New Series III/26, ISBN 0-387-50886-4, Springer-Verlag, Berlin, Germany

Nakajima H., Nunoya Y., Nozawa M., Ivano O., Takano K., Ando S. & Ohkita S. (1996). Development of High Strength Austenitic Stainless Steel for Conduit of Nb_3Al Conductor, *Advances in Cryogenic Engineering*, Vol. 42 A, pp. 323-330, ISSN 0065-2482

Porter D.A.; Easterling, K.E. & Sherif, M.Y (2009). *Phase Transformations in Metals and Alloys*, CRC Press, ISBN 978-1-4200-6210-6, NW, USA.

Ratke, L. & Vorhees, P.W. (2002). *Growth and Coarsening: Ripening in Materials*, Springer, Berlin, Germany, ISBN 3-540 – 42563-2

Sauthoff, G. (1995). *Intermetallics*, Wiley-VCH, ISBN 3-527-29320-5, Weinheim, Germany

Soriano-Vargas, O.; Saucedo-Muñoz, M.L., Lopez-Hirata, V.M. & Paniagua Mercado, A. (2010). Coarsening of β' Precipitates in an Isothermally-Aged Fe_{75}-Ni_{10}-Al_{15} Alloy, *Mater. Trans. JIM*, Vol. 51, No. x, (Month, 2010), pp.442-446, ISSN 1345-9678

Saucedo-Muñoz, M.L.;, Watanabe Y., Shoji T. & Takahashi H. (2001), Effect of Microstructure Evolution on Fracture Toughness in Isothermally Aged Austenitic Stainless Steels for Cryogenic Applications. *Journal of Cryogenic Materials*, Vol. 40, pp. 693-700. ISSN 011-2275

4

Fractal Analysis of Micro Self-Sharpening Phenomenon in Grinding with Cubic Boron Nitride (cBN) Wheels

Yoshio Ichida
Utsunomiya University,
Japan

1. Introduction

Self-sharpening phenomenon of the grain cutting edges during grinding is the main factor controlling the performance and the tool life of grinding wheels. Therefore, many studies on the relationship between the wear behavior and the self-sharpening of the grain cutting edges have been carried out (Yoshikawa, 1960; Tsuwa, 1961; Ichida et al., 1989, 1995; Malkin, 1989; Show, 1996). However, it is very difficult to evaluate this relation quantitatively because of the complexity in wear mechanism and the irregularity in shape and distribution of the grain cutting edges (Webster & Tricard, 2004). Especially, self-sharpening of the cutting edges in the grinding process with cBN wheels has not yet been clarified sufficiently (Ichida et al. 1997, 2006; Guo etal., 2007). To develop an innovative machining system using cBN grinding wheels, it is essential to clarify the self-sharpening mechanism due to the micro fracturing of the cutting edges that is the most important factor controlling the grinding ability of cBN wheel during the grinding process (Ichida et al. 2006; Kalpakjian, 1995; Comley et al., 2006).

The main purpose of this study is to evaluate quantitatively such a complicated self-sharpening phenomenon of the cutting edges in cBN grinding on the basis of fractal analysis. The changes in three-dimensional surface profile of cBN grain cutting edge in the grinding process are measured using a scanning electron microscope with four electron detectors and evaluated by means of fractal dimension.

2. Three-dimensional fractal analysis

There are several methods for calculating fractal dimension (Mandelbrot, 1983; Mandelbrot et al. 1984; Hagiwara et al., 1995; Itoh et al., 1990). In this report, we have used a 3D-fractal analysis that is expanded based on the idea in the fractal analysis using two-dimensional mesh counting method (Sakai et al., 1998). The analysis method is shown as follows. A 3D-profile under test is divided by cube grid with a mesh size of r. And then, the number of cubes intersected with 3D-profile $N(r)$ is counted. If there is a fractal nature in this 3D-profile, the relationship between $N(r)$, r and fractal dimension D_S is given by

$$N(r) = \alpha \cdot r^{-D_s} \tag{1}$$

where α is constant number.

Area of square with mesh size r is expressed r^2. Therefore, the surface area of 3D-profile $S(r)$ based on $N(r)$ is given by

$$S(r) = r^2 \cdot N(r) = \alpha \cdot r^2 \cdot r^{-D_s} \tag{2}$$

If the logarithm of both sides is taken, eq. (2) is rewritten as follows;

$$\log S(r) = \log \alpha + (2 - D_s) \log r \tag{3}$$

Fractal dimension D_S is calculated by the following equation using the proportionality constant between log $S(r)$ and log r in eq. (3).

$$D_S = 2 - \frac{d \log S(r)}{d \log r} \tag{4}$$

However, actual fractal analysis is conducted according to the following procedures by computer processing in this study. As shown in Fig. 1 (a), a square grid with mesh size r_1 is set on a 3D-profile of the top surface of grain cutting edge. It is divided to two triangular elements with mesh size r_1. Surface areas of each triangle $s_1(r_1)$ and $s_2(r_1)$ are evaluated using height coordinates in each grid point and $S(r_1)$ is decided by sum of these surface areas. Next, as shown in Fig. 1 (b), each triangle is divided with mesh size r_2 that is half a size of r_1. Surface areas of 8 triangles $s_1(r_2)$, ..., $s_8(r_2)$ are evaluated using height coordinates in 9 grid points and $S(r_2)$ is decided by sum of these surface areas. In addition, as shown in Fig. 1 (c), 8 triangles are divided with mesh size r_3 that is half a size of r_2. Surface areas of 32 triangles $s_1(r_3)$, ..., $S_{32}(r_3)$ are evaluated using height coordinates in 25 grid points and $S(r_3)$ is decided by sum of these surface areas. Afterward, mesh size r is scaled down and the surface area of 3D-profile is evaluated as follows;

$$S(r_n) = \sum_{i=1}^{2^{2n-1}} s_i(r_n) \tag{5}$$

On the basis of these equations, r is taken on the horizontal log axis, and $S(r)$ is taken on the vertical log axis. When data points are on a straight line in double log plot, fractal dimension D_S is given by a slope of the straight line. Figure 2 shows an example of relationship between $S(r)$ and r (Sample: surface profile of cBN cutting edge shown in Fig.1). Fractal nature is approved in a region of $0.4 < r < 4$ μm. From a slope of the straight line, it is decided that fractal dimension is 2.015.

3. Three-dimensional observation of wheel working surface

3.1 Experimental procedure

Grinding experiments were conducted with surface plunge grinding method on a horizontal spindle surface grinding machine. The schematic illustration of the experimental setup is shown in Fig. 3. A vitrified cBN wheel with a replaceable cBN segment shown in Fig. 3 was used to observe directly the profile of the wheel surface in the grinding process using a three-dimensional (3D) scanning electron microscope with four electron probes (3D-SEM/

EDM3000) (Ichida, 2008). The cBN segment detached for the observation can be precisely returned to former state again. It was confirmed experimentally that the cBN wheel with the replaceable cBN segment has almost the same grinding ability as the usual complete cBN wheel (Fujimoto, 2006).

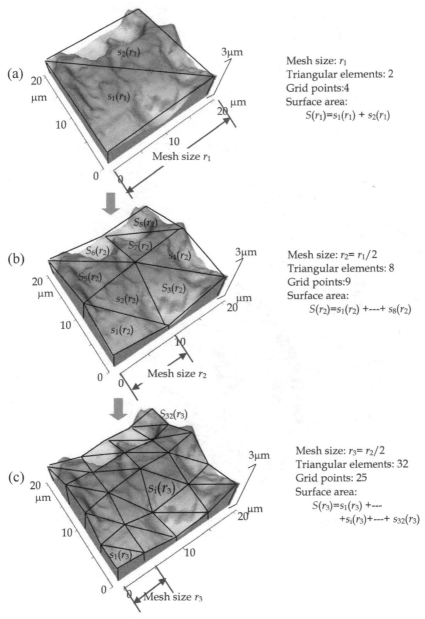

Fig. 1. Method of 3D-fractal analysis (Sample: surface profile of cBN abrasive grain).

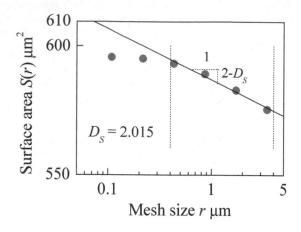

Fig. 2. Relationship between surface area $S(r)$ and mesh size r
(Sample: surface profile shown in Fig.1).

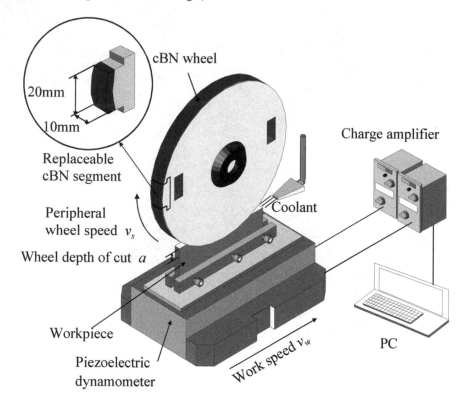

Fig. 3. Schematic illustration of the experimental setup.

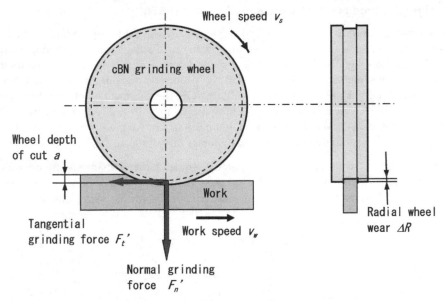

Fig. 4. Measurements of grinding characteristic palameters.

The expermental conditions are listed in Table 1. Representative single crystal cBN grain was used for cBN wheel. The dressing of cBN wheel was performed using a rotary diamond dresser (Dressing wheel: SD40Q75M) equipped with an AE sensor under the following dressing conditions: peripheral dressing speed 16.5 m/s, peripheral wheel speed ratio 0.5, dressing lead 0.1 mm/rev, dressing depth of cut 2μm×5 times. High speed steel SKH51/JIS (M2/ASTM) is used as the workpiece material.

Grinding method	Surface plunge grinding(Up-cut)
Grinding wheel	CBN80L100V Dimensions:200 ^{D}× 10^T [mm]
cBN grain	Single crystal cBN
Peripheral wheel speed v_s	33 [m/s]
Work speed v_w	0.1 [m/s]
Wheel depth of cut a	10 [μm]
Grinding fluid	Soluble type (JIS W-2-2) 2% dilution
Workpiece	High speed steel (JIS/SKH51) Hardness: 65HRC Dimensions:100^l×5^t×30^h [mm]

Table 1. Grinding conditions.

3.2 Measuring method of wheel surface profile with 3D-SEM

This 4-channel secondary electron (SE) detection system enables quantitative surface roughness measurements and enhances the topography by displaying the differential signal calculated from the 4 signals. The intensities of these detected signals are determined by the tilt angle of the specimen surface in relation to the geometric positioning of the detectors. The quantitative angular information can be obtained by the subtraction between the signal intensities of the detectors. By calculating 4 tilting angles (two in X-direction and two in Y-direction) on many spots in the X-Y matrix taken on the specimen, the surface topography of the specimen can be accurately re-constructed by integrating these angles over the matrix.

In this system, no eucentric tiling for stereo pairs is required, thereby simplifying operation and allowing much better precision and resolution than conventional SEMs using stereo photogrammetry. The vertical resolution in measuring a 3D profile using this 3D-SEM is 1 nm.

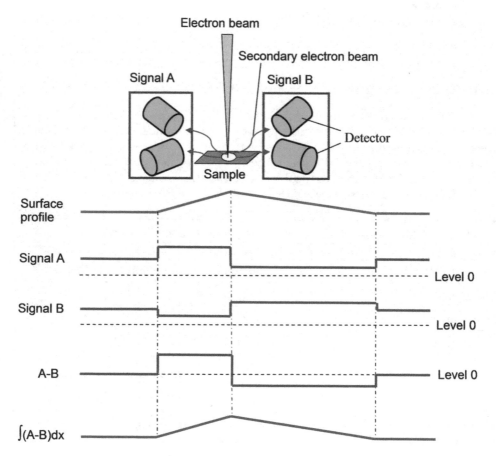

Fig. 5. Illustration of the 4-channel SE detector layout detailing the measurement principle of 3D-SEM.

4. Grinding wheel wear and wheel working surface

Grinding wheel wear is an important consideration because adversely affects the shape and accuracy of ground surface. Grinding wheel wear by three different mechanisms: attritious grain wear, grain fracture and, and bond fracture, as shown in Fig.6. In attritious wear, the cutting edges of a sharp grain dull by attrition, developing a wear flat. Wear is caused by the interaction of the grain with the workpiece material, involving both physical and chemical reactions. These reactions are complex and involve diffusion, chemical degradation or decomposition of the grain, fracture at a microscopic scale, plastic deformation, and melting. If the wear flat caused by attritious wear is excessive, the grain becomes dull and grinding becomes inefficient and produces undesirable high temperatures. Optimally, the grain should fracture or fragment at a moderate rate, so that new sharp cutting edges are produced continuously during grinding. This phenomenon is self-sharpening. However, self-sharpening by a large fracture is not suitable for precision grinding, because it gives large wheel wear and bad surface roughness during grinding. Therefore, self-sharpening due to micro fracture as shown in Fig.6 is suitable for effective precision grinding, because it offers small wheel wear and good surface roughness. We call this phenomenon 'micro self sharpening'(Ichida, 2008).

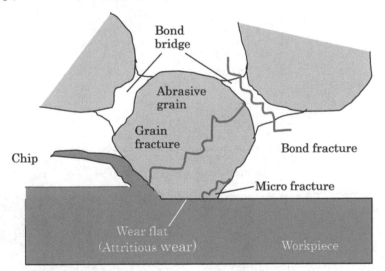

Fig. 6. Wear mechanisms of abrasive grain during grinding.

Fig. 7 shows the change in radial wheel wear ΔR with increasing the accumulated stock removal (cumulative volume of material removed per unit grinding width) V_w' when grinding under the conditions indicated in Table 1. At the same time, some typical sequential SEM images of the wheel working surface with an increase of stock removal are shown in this figure. The wear process of grinding wheel can be divided into two different regions: a) initial wear region over stock removal range from 0 to 1000 mm³/mm, in which a rapid increase of wheel wear occurs with increasing stock removal, b) steady-state wear region over stock removal range larger than 1000 mm³/mm, in which the wheel wear rate maintains a nearly constant value. In the initial wear region, a releasing of grain due to bond fracture and grain fracture are sometimes observed, as shown in grains C, D and so on. However, they are not observed so much in the steady-state wear region. The wheel wear in steady-state region dominantly occurs due to attritious wear and micro fracture, as shown in grains A, B and so on. As a typical example, a high magnification SEM image of grain A is shown in Fig.8 (a). Fig.8 (b) is its contour map. Wear flat developed due to attritiou wear and some brittle surfaces generated by micro fracture can be observed on the point of the grain.

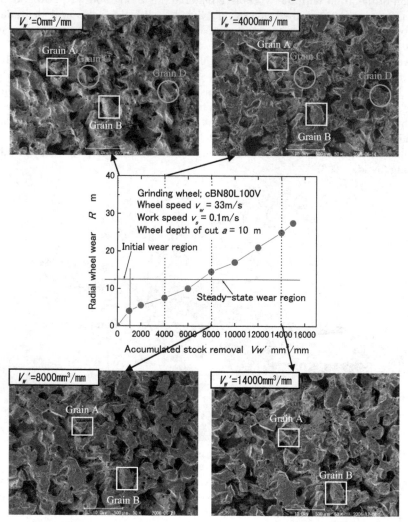

Fig. 7. Change of radial wheel wear with increasing stock removal and typical sequential SEM images of wheel working suraface.

There is little research that has quantitatively evaluated self-sharpening phenomenon of grinding wheel. We have tried to grasp the actual behavior of self-sharpening and evaluate it by the attritious wear flat area percentage of the grain cutting edge (Ichida, 2008, 2009).

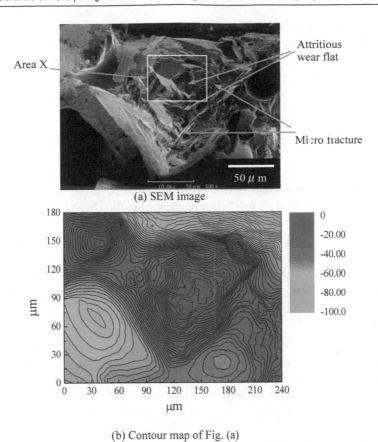

(a) SEM image

(b) Contour map of Fig. (a)

Fig. 8. High magnification SEM image and its contour map of grain A in Fig.7 (V_w'=4 000mm³/mm).

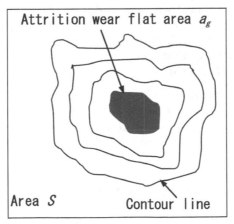

Fig. 9. Measuring method of attritious wear flat percentage.

Figure 9 shows the measuring method of the attritious wear flat percentage. Attritious wear flat area a_g in the area S to be observed is measured using SEM image and contour map made by 3D-profiles. Here, attritious wear flat percentage A_g is given by:

$$A_g = \frac{a_g}{S} \times 100[\%] \tag{6}$$

5. Self-sharpening phenomenon due to micro fracture of cutting edges

Grain cutting edges on the wheel surface change their shapes in various forms with the progress of wheel wear when the accumulated stock removal V_w' increases. Many sequential observations of the grain cutting edge with accumulated stock removal have been carried out using 3D-SEM. The typical result is shown in Fig. 10. Those are high magnification images of area X on grain A in Fig. 8(a).

The surface with a micro unevenness formed by the diamond dresser is observed on the tip of the grain cutting edge after dressing, as shown in Fig. 10 (a). And after grinding the stock removal V'_w = 500 mm³/mm, an attritious wear flat is observed in the center part on the top surface of the grain cutting edge, as shown in Fig. 10 (b). Moreover, at V'_w = 2000 mm³/mm, the wear flat becomes larger than that at V'_w = 500 mm³/mm, as indicated in the comparison between Figs. 10 (b) and (c). The ductile attritious wear flat area takes the largest value at V'_w = 2000 mm³/mm in the grinding process, as seen from all SEM images in Fig.4. However, between the stock removals from 2000 to 4000 mm³/mm, some micro fractures take place at the lower left side part of cutting edge and consequently the wear flat area decreases a little, as indicated in the comparison between Figs. 10(c) and (d). Moreover,between the stock removals from 4000 to 10000 mm³/mm, as many micro fractures take place repeatedly, the ductile attritious wear flat area is decreased and some new sharp edges are formed on the top surface of cutting edge, as shown in the comparison between Figs. 10 (e) and (f).

In addition, between the stock removals from 10000 to 12000 mm³/mm, a small fracture takes place at the right side part of cutting edge and some new sharp edges are formed, and at the same time the wear flat is formed slightly in the center part on the cutting edge surface, as indicated in the comparison between Figs. 10 (f) and (g). Afterward, between the stock removals from 12000 to 14000 mm³/mm, some new sharp edges due to the micro fracture are observed in the middle part on the cutting edge top surface, while the new attritious wear flat is formed again at the upper part of the cutting edge, as indicated in the comparison between Figs. 10 (g) and (h).

Thus, although the grain cutting edges become dull due to the ductile attritious wear, they can reproduce and maintain their sharpness due to the micro fractures occurred repeatedly on their top surfaces. Namely, an actual behavior of the self-sharpening phenomenon due to the micro fracture may be grasped on the basis on the sequential SEM observation method used in this study.

6. Evaluation of self-sharpening using fractal dimension

As mentioned above, the shape of the cutting edges on the wheel working surface is variously changed due to the fracture wear or the attritious wear when the accumulated stock removal increases. Such a complicated wear process is evaluated using 3D-fractal dimension. Fractal dimension is calculated in an area of 27.4×20.6 μm² enclosed with white frame in Fig. 10. The center of these areas is almost located in the top part of the cutting edge that acts as an effective edge. The range of mesh size r is $0.11 < r < 6.4$ μm.

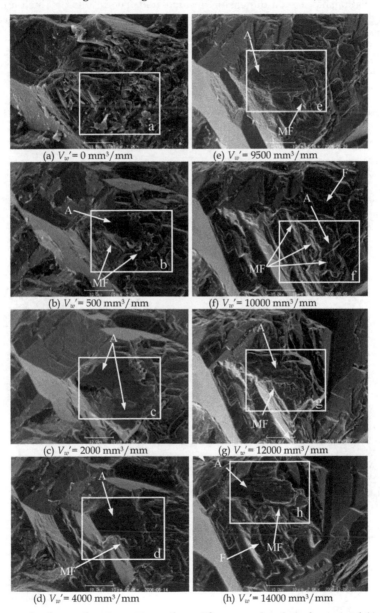

(a) $V_w' = 0$ mm³/mm

(b) $V_w' = 500$ mm³/mm

(c) $V_w' = 2000$ mm³/mm

(d) $V_w' = 4000$ mm³/mm

(e) $V_w' = 9500$ mm³/mm

(f) $V_w' = 10000$ mm³/mm

(g) $V_w' = 12000$ mm³/mm

(h) $V_w' = 14000$ mm³/mm

Fig. 10. Change in shape of grain cutting edge with accumulated stock removal (Area X on grain A in Fig.8(a)) (A: attritious wear, MF: micro fracture, F: fracture).

Figure 11 shows the 3D-profiles of the typical eight areas a, b, ----, h on the cutting edge used for fractal analysis (areas enclosed with white frame in Fig. 10). The relationships between mesh size r and surface area $S(r)$ obtained in these typical eight areas are shown in Fig. 12. This figure indicates that the fractal nature is approved in a region of $0.4 < r < 4$ μm. Using this relationship, fractal dimension is calculated. The results obtained are shown in Fig. 13. The fractal dimension changes complexly and randomly when the accumulated stock removal increases.

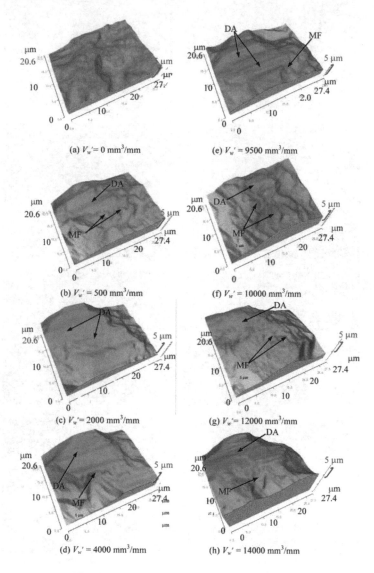

Fig. 11. Sequential 3D-profiles of typical area on grain cutting edge used for fractal analysis (DA: ductile attritious wear, MF: micro fracture).

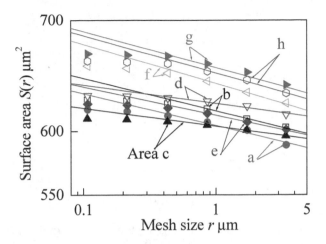

Fig. 12. Relationship between surface area $S(r)$ and mesh size r in areas shown in Fig.11.

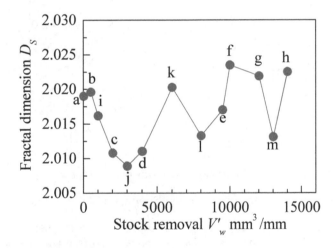

Fig. 13. Change in fractal dimension on top surface profile of grain cutting edge with accumulated stock removal.

To consider the reason for such complicated change of fractal dimension, the attritious wear flat area percentage of the cutting edge was measured. In this study, a percentage of ductile attritious wear area in same area of 27.4 × 20.6 μm² used for fractal analysis is measured and defined as attritious wear flat area percentage A_g. Figure 14 shows the change in the attritious wear flat area percentage A_g of the grain cutting edge with increasing accumulated stock removal.

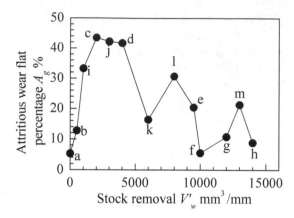

Fig. 14. Change in attritious wear flat percentage on top surface of grain cutting edge with accumulated stock removal.

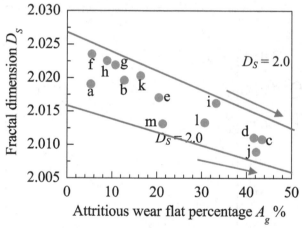

Fig. 15. Relationship between fractal dimension D_S and attritious wear flat percentage A_g

As shown in Figs. 13 and 14, the cutting edge after dressing comparatively takes a high fractal dimension (D_s =2.02), because it has complicated surface with a micro ruggedness formed by the diamond dresser. And then, between the stock removals from 500 to 3000mm³/mm, fractal dimension decreases because the attritious wear flat increases with the accumulated stock removal. In addition, between the stock removals from 2000 to 4000 mm³/mm, fractal dimension indicates the lowest value (D_s =2.01) because the attritious wear flat takes the highest value. Afterward, over a range of stock removals from 4000 to 6000 mm³/mm, fractal dimension tends to increase because the attritious wear flat decreases and new sharp cutting edges are formed by self-sharpening due to micro fractures. However, between the stock removals from 6000 to 8000 mm³/mm, fractal dimension decreases slightly because of a little increase in attritious wear flat area. Moreover, fractal dimension increases rapidly over a range of stock removals from 8000 to 10000 mm³/mm because the attritious wear flat area decreases and micro fracture occurs repeatedly, i.e., self-sharpening due to micro fracture takes place actively. Afterward, although the fractal dimension decreases because of increasing in attritious wear flat at the stock removal 13000 mm³/mm, it increases again because self-sharpening due to micro fracture takes place actively over a range of stock removals from 13000 to 14000 mm³/mm.

As mentioned above, self-sharpening of the grain cutting edge can be characterized using fractal dimension. Especially, these results show that there is a close relationship between fractal dimension D_S and attritious wear flat percentage A_g. Figure 15 shows relationship between fractal dimension and attritious wear flat percentage. The alphabets in Fig.15 correspond to those in Figs.10, 11, 12, 13 and 14. As shown in this figure, fractal dimension decreases with increasing the attritious wear flat percentage and then becomes 2.0 at A_g = 100 % (perfect smooth surface) as a limit value. Thus, there is a negative correlation between fractal dimension and attritious wear flat percentage.

7. Effect of self-sharpening on grinding characteristics

Fig.16 and 17 show the changes of grinding forces and ground surface roughness with increasing accumulated stock removal, respectively. Under this grinding condition, grinding forces maintains a stable level in the steady-state wear region. Especially, tangential grinding force keeps a small variation between 4 and 6 N/mm in this wear region. On the other hand, although surface roughness increases with increasing stock removal, its increasing rate maintains comparatively low level. Thus, high grinding ability of cBN wheel is brought from such self-sharpening due to micro fracture of grain cutting edges, that is, micro self-sharpening phenomenon.

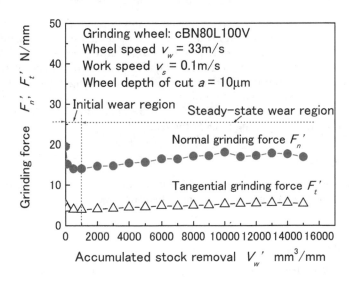

Fig. 16. Changes in grinding forces with increasing accumulated stock removal.

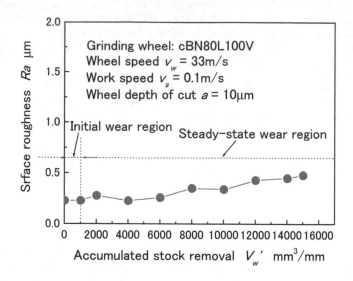

Fig. 17. Change of ground surface roughness with increasing accumulated stock removal.

8. Conclusions

The changes in three-dimensional surface profile of grain cutting edge in the grinding process with cBN wheels have been measured using a 3D-SEM and evaluated by means of fractal dimension. The main results obtained in this study are summarized as follows;

1. Actual behavior of self-sharpening phenomenon due to the micro fracture in the grinding process can be grasped using sequential observation method with 3D-SEM.
2. The fractal dimension for surface profile of the cutting edge formed by the micro fracture is higher than that of the cutting edge formed due to ductile attitious wear. An increase in ductile attritous wear flat area on the grain cutting edge results in a decrease in fractal dimension for its surface profile.
3. The complicated changes in shape of the cutting edge due to self-sharpening can be evaluated quantitatively using fractal dimension.

9. Acknowledgment

This research was supported in part by Grants-in-Aid for General Science Research (C) (No.19560106) from the Ministry of Education, Culture, Sports, Science and Technology of Japan.

10. References

Yoshikawa, H. (1960). Process of Wear in Grinding Wheel with Fracture of Bond and Grain, *Journal of the Japan Society for Precision Engineering*, Vol.26, No. 11, (1960), pp.691-700, ISSN0912-0289

Tsuwa, H. (1961). On the Behaviors of Abrasive Grains in Grinding Process (Part 4)-Microscopic Observations of Cutting Edges-, *Journal of the Japan Society for Precision Engineering*, Vol. 27, No. 11,(1961), pp. 719-725 , ISSN0912-0289

Kalpakjian, S. (1995), *Manufacturing Engineering and Technology*, Third Edition, Addison-Wesley Publishing Company Inc. , ISBN 0-201-53846-6, New York, pp.795-798.

Show, M. C. (1996), *Principles of Abrasive Processing*, Clarendon Press, Oxford, ISBN 0-19-859021-0, pp.55-62.

Malkin, S. (1989). *Grinding Technology: Theory and Applications of Machining with Abrasives*, Ellis Horwood Limited, Chichester, UK, PP.197-202. ISBN 0-85312-756-5

Ichida, Y.; Fredj, N. B. & Usui, N. (1995). The Micro Fracture Wear of Cutting Edges in CBN Grinding, *The Second International ABTEC Conference*, Vol. 11, (1995), pp. 501-504.

Ichida, Y.; Kishi, K.; Suyama, Y. & Okubo, J. (1989). Study of Creep Feed Grinding with CBN Wheels, -Characteristics of Wheel Wear-, Vol.55, No.8, pp.1468-1474, ISSN0912-0289

Ichida, Y. & Kishi, K. (1997). The Development of Nanocrystalline cBN for Enhanced Superalloy Grinding Performance, *Transactions of the ASME, Journal of Manufacturing, Science and Engineering*, Vol. 119, No. 1, (1997), pp.110-117. ISSN 1087-1357

Webster, J. & Tricard, M. (2004). Innovations in Abrasive Production for Precision Grinding, *Annals of the CIRP*, Vol.53, No.2, pp.597-617, ISSN 0007-8506

Ichida, Y.; Sato, R.: Morimoto, Y. & Inoue, Y. (2006). Profile Grinding of Superalloys with Ultrafine-Crystalline cBN Wheels, *JSME International Journal, Series C*, Vol.49, No.1, pp.94-99, ISSN 1344-7653.

Guo, C.; Shi, Z.; Attia, H. & McIntosh, D. (2007). Power and Wheel Wear for Grinding Nickel Alloy with Plated CBN Wheels, *Annals of the CIRP*, Vol.56, No.1, pp.343-346, ISSN 0007-8506

Comley, P.; Walton, I.; Jin, T. & Stephenson, D. J. (2006). A High Material Removal Rate Grinding Process for the Production of Automotive Crankshafts, *Annals of the CIRP*, Vol.55, No.1, pp.347-350, ISSN 0007-8506

Mandelbrot, B. B. (1983). The Fractal Geometry of Nature, Freeman, W. H. and Company, New York, (1983), pp.109-111, ISBN4-532-06254-3

Mandelbrot, B. B.; Passoja, D. E. & Paullay, A. J. (1984). Fractal Characterization of Fracture Surfaces of Metals, *Nature*, Vol. 308, pp.1571-1572, ISSN 0028-0836.

Hagiwara, S.; Obikawa, T. & Yanai, H. (1995). Evaluation of Lapping Grains Based on Shape Characteristics, *Journal of the Japan Society for Precision Engineering*, Vol. 61, No.12, pp.1760-1764, ISSN0912-0289

Itoh, N.; Tsukada, T.& Sasajima, K. (1990). Three-Dimensional Characerization of Engineering Surface by Fractal Dimension, *Bulletin of the Japan Society for Prcision Engineering*, Vol.24, No.2, pp.148-149.

Sakai, T.; Sakai, T. & Ueno, A. (1998). Fractal Analysis of Metal Surface Mechanically Finished by Several Methods, *Transactions of the Japan Society of Mechanical Engineers, Series A*, Vol. 64, No. 620, (1998-4), pp.1104-1112, ISSN 1884-8338.

Fujimoto, M.; Ichida, Y.; Sato, R. & Morimoto, Y. (2006), Characterization of Wheel Surface Topography in cBN Grinding, *JSME International Journal, Series C*, Vol.49, No.1, pp.106-113, ISSN 1344-7653.

Ichida, Y.; Sato, R.; Fujimoto, M. & Tanaka, H. (2008), Fractal Analysis of Grain Cutting Edge Wear in Superabrasive Grinding, *JSME Journal of Advanced Mechanical Design, Systems, and Manufacturing*, Vol.2, No.4, pp.640-650 ISSN 1881-3054 .

Ichida, Y.; Sato, R.; Fujimoto, M. & Fredj, N. B. (2009). Fractal Analysis of Self-Sharpening Phenomenon in cBN Grinding, *Key Engineering*, Vols. 389-390, (2009), pp.42-47, ISBN-13978-0-87849-364-7

Ichida, Y.; Fujimoto, M.; Akbari, J. & Sato, R. (2008). Evaluation of Cutting Edge wear in cBN Grinding Based on Fractal Analysis, *6th International Scientific and technical Symposium on Manufacturing and Materials*, Monastir, Tunisia, pp.287-294.

5

Strength and Microstructure of Cement Stabilized Clay

Suksun Horpibulsuk

Suranaree University of Technology,
Thailand

1. Introduction

The soil/ground improvement by cement is an economical and worldwide method for pavement and earth structure works. Stabilization begins by mixing the in-situ soil in a relatively dry state with cement and water specified for compaction. The soil, in the presence of moisture and a cementing agent becomes a modified soil, i.e, particles group together because of physical-chemical interactions among soil, cement and water. Because this occurs at the particle level, it is not possible to get a homogeneous mass with the desired strength. Compaction is needed to make soil particles slip over each other and move into a densely packed state. In this state, the soil particles can be welded by chemical (cementation) bonds and become an engineering material (Horpibulsuk et al., 2006). To reduce the cost of ground improvement, the replacement of the cement by waste materials such as fly ash and biomass ash is one of the best alternative ways. In many countries, the generation of these waste materials is general far in excess of their utilization. A feasibility study of utilizing these ashes (waste materials) to partially replace Type I Portland cement is thus interesting.

The effects of some influential factors, i.e., water content, cement content, curing time, and compaction energy on the laboratory engineering characteristics of cement-stabilized soils have been extensively researched (Clough et al., 1981; Kamon & Bergado, 1992; Yin & Lai, 1998; Miura et al., 2001; Horpibulsuk & Miura, 2001; Horpibulsuk et al., 2003, 2004a, 2004b, 2005, 2006, 2011a). The field mixing effect such as installation rate, water/cement ratio and curing condition on the strength development of cemented soil was investigated by Nishida et al. (1996) and Horpibulsuk et al. (2004c, 2006 and 2011b). Based on the available compression and shear test results, many constitutive models were developed to describe the engineering behavior of cemented clay (Gens and Nova, 1993; Kasama et al., 2000; Horpibulsuk et al., 2010a; Suebsuk et al., 2010 and 2011). These investigations have mainly focused on the mechanical behavior that is mainly controlled by the microstructure. The structure is fabric that is the arrangement of the particles, clusters and pore spaces in the soil as well as cementation (Mitchell, 1993). It is thus vital to understand the changes in engineering properties that result from the changes in the influential factors.

This chapter attempts to illustrate the microstructural changes in cement-stabilized clay to explain the different strength development according to the influential factors, i.e., cement content, clay water content, fly ash content and curing time. The unconfined compressive strength was used as a practical indicator to investigate the strength development. The microstructural analyses were performed using a scanning electron microscope (SEM), mercury intrusion porosimetry (MIP), and thermal gravity (TG) tests. For SEM, the cement stabilized samples were broken from the center into small fragments. The SEM samples were frozen at -195°C by immersion in liquid nitrogen for 5 minutes and evacuated at a pressure of 0.5 Pa at -40°C for 5 days (Miura et al., 1999; and Yamadera, 1999). All samples were coated with gold before SEM (JEOL JSM-6400) analysis.

Measurement on pore size distribution of the samples was carried out using mercury intrusion porosimeter (MIP) with a pressure range from 0 to 288 MPa, capable of measuring pore size diameter down to 5.7 nm (0.0057 micron). The MIP samples were obtained by carefully breaking the stabilized samples with a chisel. The representative samples of 3-6 mm pieces weighing between 1.0-1.5 g were taken from the middle of the cemented samples. Hydration of the samples was stopped by freezing and drying, as prepared in the SEM examination. Mercury porosimetry is expressed by the Washburn equation (Washburn, 1921). A constant contact angle (θ) of 140° and a constant surface tension of mercury (γ) of 480 dynes/cm were used for pore size calculation as suggested by Eq.(1)

$$D = -(4\gamma\cos\theta) / P \qquad (1)$$

where D is the pore diameter (micron) and P is the applied pressure (MPa).

Thermal gravity (TG) analysis is one of the widely accepted methods for determination of hydration products, which are crystalline $Ca(OH)_2$, CSH, CAH, and CASH, ettringite (Aft phases), and so on (Midgley, 1979). The CSH, CAH, and CASH are regarded as cementitious products. $Ca(OH)_2$ content was determined based on the weight loss between 450 and 580°C (El-Jazairi and Illston, 1977 and 1980; and Wang et al., 2004) and expressed as a percentage by weight of ignited sample. When heating the samples at temperature between 450 and 580°C, $Ca(OH)_2$ is decomposed into calcium oxide (CaO) and water as in Eq. (2).

$$Ca(OH)_2 \text{-----------------------} > CaO + H_2O \qquad (2)$$

Due to the heat, the water is lost, leading to the decrease in overall weight. The amount of $Ca(OH)_2$ can be approximated from this lost water by Equation (2), which is 4.11 times the amount of lost water (El-Jazairi and Illston, 1977 and 1980). The change of the cementitious products can be expressed by the change of $Ca(OH)_2$ since they are the hydration products.

2. Compaction and strength characteristics of cement stabilized clay

Compaction characteristics of cement stabilized clay are shown in Figure 1. The clay was collected from the Suranaree University of Technology campus in Nakhon Ratchasima, Thailand. It is composed of 2% sand, 45% silt and 53% clay. Its specific gravity is 2.74. The liquid and plastic limits are approximately 74% and 27%, respectively. Based on the Unified Soil Classification System (USCS), the clay is classified as high plasticity (CH). It is found that the maximum dry unit weight of the stabilized samples is higher than that of the unstabilized samples whereas their optimum water content is practically the same. This characteristic is the same as that of cement stabilized coarse-grained soils as reported by Horpibulsuk et al. (2006). The adsorption of Ca^{2+} ions onto the clay particle surface decreases the repulsion between successive diffused double layers and increases edge-to-face contacts between successive clay sheets. Thus, clay particles flocculate into larger clusters, which increases in the plastic limit with an insignificant change in the liquid limit (*vide* Table 1). As such, the plasticity index of the mixture decreases due to the significant increase in the plastic limit. Because the *OWC* of low swelling clays is mainly controlled by the liquid limit (Horpibulsuk et al., 2008 and 2009), the *OWC*s of the unstabilized and the stabilized samples are almost the same (*vide* Table 1). Figure 2 shows the compaction curve of the fly ash (FA) blended cement stabilized clay for different replacement ratios (ratios of cement to fly ash, C:F) compared with that of the unstabilized clay. Two fly ashes are presented in the figure: original, OFA (D_{50} = 0.03 mm) and classified, CFA (D_{50} = 0.009 mm) fly ashes. It is noted that the compaction curve of the stabilized clay is insignificantly dependent upon replacement ratio and fly ash particles. Maximum dry unit weight of the stabilized clay is higher than that of the unstabilized clay whereas their optimum water content is practically the same.

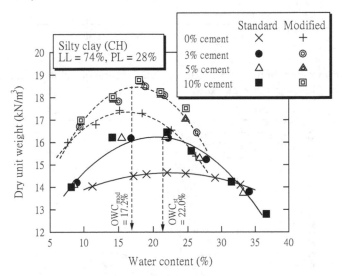

Fig. 1. Plots of dry unit weight versus water content of the uncemented and the cemented samples compacted under standard and modified Proctor energies (Horpibulsuk et al., 2010b)

Cement (%)	Atterberg's limits (%)			OWC (%)		γ_{dmax} (kN/m³)	
	LL	PL	PI	Std.	Mod.	Std.	Mod.
0	74.1	27.5	46.6	22.4	17.2	14.6	17.4
3	74.1	45.0	29.1	22.2	17.5	16.2	18.5
5	72.5	45.0	27.5	21.8	17.3	16.2	18.7
10	71.0	44.8	26.2	22.0	17.4	16.4	18.8

Table 1. Basic properties of the cemented samples

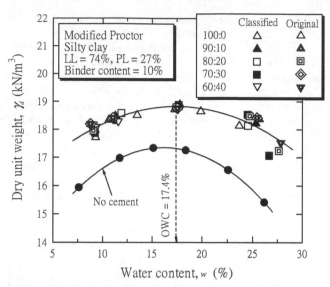

Fig. 2. Compaction curves of the OFA and CFA blended cement stabilized clay and the unstabilized clay (Horpibulsuk et al., 2009)

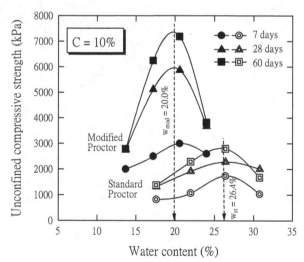

Fig. 3. Effect of compaction energy and curing time on strength development (Horpibulsuk et al., 2010b)

Typical strength-water content relationships for different curing times and compacton energies of the stabilized samples are shown in Figure 3. The strength of the stabilized samples increases with water content up to 1.2 times the optimum water content and decreases when the water content is on the wet side of optimum. At a particular curing time, the strength curve depends on the compaction energy. As the compaction energy increases, the maximum strength increases and the water content at maximum strength decreases. For the same compaction energy, the strength curves follow the same pattern for all curing times, which are almost symmetrical around 1.2*OWC* for the range of the water content tested. Figure 4 shows the strength versus water content relationship of the CFA blended cement stabilized clay at different replacement ratios after 60 days of curing compared with that of the unstabilized clay. The maximum strengths of the stabilized clay are at about 1.2*OWC* whereas the maximum strength of the unstabilized clay is at *OWC* (maximum dry unit weight). This is because engineering properties of unstabilized clay are mainly dependent upon the densification (packing).

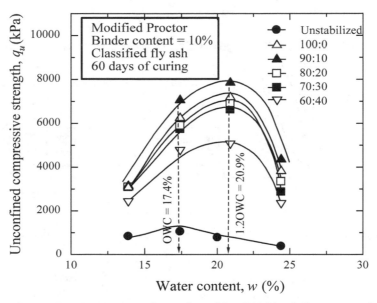

Fig. 4. Strength versus water content relationship of the CFA blended cement stabilized clay at different replacement ratios and 60 days of curing (Horpibulsuk et al., 2009)

Figure 5 shows the strength development with cement content (varied over a wide range) of the stabilized samples compacted under the modified Proctor energy at 1.2 *OWC* (20%) after 7 days of curing. The strength increase can be classified into three zones. As the cement content increases, the cement per grain contact point increases and, upon hardening, imparts a commensurate amount of bonding at the contact points. This zone is designated as the *active zone*. Beyond this zone, the strength development slows down while still gradually increasing. The incremental gradient becomes nearly zero and does not make any further significant improvement. This zone is referred to as the *inert zone* (*C* = 11-30%). The strength decrease appears when *C* > 30%. This zone is identified as the *deterioration* zone.

Influence of replacement ratio on the strength development of the blended cement stabilized clay compacted at water content (*w*) of 1.2*OWC* (*w* = 20.9%) for the five curing times is presented in Figure 6. For all curing times, the samples with 20% replacement ratio exhibit almost the same strength as those with 0% replacement ratio. The 30 and 40% replacement samples exhibit lower strength than 0% replacement samples. The samples with 10% replacement ratio exhibit the highest strength since early curing time. The sudden strength development with time is not found for all replacement ratios. This finding is different from concrete technology where the role of fly ash as a pozzolanic material comes into play after a long curing time (generally after 60 days). In other words, the strength of concrete mixed with fly ash is higher than that without fly ash after about 60 days of curing.

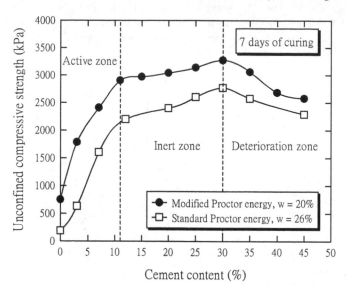

Fig. 5. Strength development as a function of cement content (Horpibulsuk et al., 2010b)

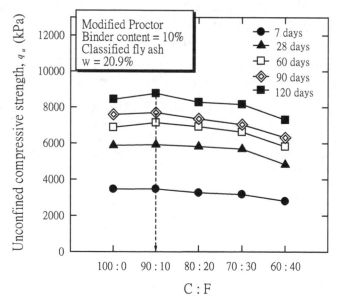

Fig. 6. Relationship between strength development and replacement ratio of the CFA blended cement stabilized clay at different curing times (Horpibulsuk et al., 2009)

3. Microstructure of cement stabilized clay

3.1 Unstabilized clay

For compacted fine-grained soils, the soil structure mainly controls the strength and resistance to deformation, which is governed by compaction energy and water content. Compaction breaks down the large clay clusters into smaller clusters and reduces the pore space. Figure 7 shows SEM photos of the unstabilized samples compacted under the modified Proctor energy at water contents in the range of 0.8OWC to 1.2OWC. On the wet side of optimum (*vide* Figure 7c), a dispersed structure is likely to develop because the quantity of pore water is enough to develop a complete double layer of the ions that are attracted to the clay particles. As such, the clay particles and clay clusters easily slide over each other when sheared, which causes low strength and stiffness. On the dry side of optimum (*vide* Figure 7a), there is not sufficient water to develop a complete double-layer; thus, the distance between two clay platelets is small enough for van der Waals type attraction to dominate. Such an attraction leads to flocculation with more surface to edge bonds; thus, more aggregates of platelets lead to compressible flocs, which make up the overall structure. At the *OWC*, the structure results from a combination of these two characteristics. Under this condition, the compacted sample exhibits the highest strength and stiffness.

(a) $w = 14\%$ (0.8OWC) (b) $w = 17\%$ (OWC)

(c) $w = 20\%$ (1.2OWC)

Fig. 7. SEM photos of the uncemented samples compacted at different molding water contents under modified Proctor energy (Horpibulsuk et al., 2010b)

3.2 Stabilized clay

3.2.1 Effect of curing time

Figure 8 shows SEM photos of the 10% cement samples compacted at w = 20% (1.2OWC) under the modified Proctor energy and cured for different curing times. After 4 hours of curing, the soil clusters and the pores are covered and filled by the cement gel (hydrated cement) (*vide* Figure 8a). Over time, the hydration products in the pores are clearly seen and the soil-cement clusters tend to be larger (*vide* Figures 8b through d) because of the growth of cementitious products over time (*vide* Table 2).

The effect of curing time on the pore size distribution of the stabilized samples is illustrated in Figure 9. It is found that, during the early stage of hydration (fewer than 7 days of curing), the volume of pores smaller than 0.1 micron significantly decreases while the volume of pores larger than 0.1 micron slightly increases. This result shows that during 7 days of curing, the cementitious products fill pores smaller than 0.1 micron and the coarse particles (unhydrated cement particles) cause large soil-cement clusters and large pore space. After 7 days of curing, the volume of pores larger than 0.1 micron tends to decrease while the volume of pores smaller than 0.1 micron tends to increase possibly because the cementitious products fill the large pores (larger than 0.1 micron). As a result, the volume of small pores (smaller than 0.1 micron) increases, and the total pore volume decreases.

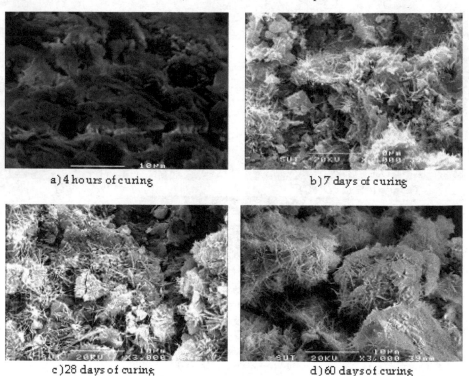

a) 4 hours of curing   b) 7 days of curing

c) 28 days of curing   d) 60 days of curing

Fig. 8. SEM photos of the 10% cement samples compacted at 1.2OWC under modified Proctor energy at different curing times (Horpibulsuk et al., 2010b)

Curing time (days)	Weight loss (%)	Ca(OH)$_2$ (%)
7	1.52	6.25
28	1.65	6.78
60	1.85	7.63

Table 2. Ca(OH)$_2$ of the 10% cement samples compacted at 1.2OWC at different curing times under modified Proctor energy (Horpibulsuk et al., 2010b)

3.2.2 Effect of cement content

Figures 10 and 11 and Table 3 show the SEM photos, pore size distribution, and the amount of Ca(OH)$_2$ of the stabilized samples compacted at w = 20% under the modified Proctor energy for different cement contents after 7 days of curing. Figures 10a-c, 10d-g, and 10h-j show SEM photos of the cemented samples in the active, inert, and deterioration zones, respectively. The SEM photo of the 3% cement sample (Figure 10a) is similar to that of the unstabilized sample because the input of cement is insignificant compared to the soil mass. As the cement content increases in the active zone, hydration products are clearly seen in the pores (*vide* Figures 10b and c) and the cementitious products significantly increase (Table 3).

The cementitious products not only enhance the inter-cluster bonding strength but also fill the pore space, as shown in Figure 11: the volume of pores smaller than 0.1 micron is significantly reduced with cement, thus, the reduction in total pore volume. As a result, the strength significantly increases with cement. For the inert zone, the presence of hydration products (Figures 10d to g) and cementitious products (Table 3) is almost the same for 15-30% cement. This results in an insignificant change in the pore size distribution and, thus, the strength. For the deterioration zone (Figure 10h-j), few hydration products are detected. Both the volumes of the highest pore size interval (1.0-0.1 micron pores) and the total pore tend to increase with cement (Figure 11). This is because the increase in cement content significantly reduces the water content, which decreases the degree of hydration and, thus, cementitious products (Table 3).

3.2.3 Effect of fly ash

Figures 12 and 13 show SEM photos of the CFA blended cement stabilized clay compacted at w = 1.2OWC (w = 20.9%) and cured for 28 and 60 days at different replacement ratios. The fly ash particles are clearly shown among clay-cement clusters especially for 30% replacement ratio (C:F = 70:30) for both curing times (Figures 12a and 13a). It is noted that the hydration products growing from the cement grains connect fly ash particles and clay-cement clusters together. Some of the surfaces of fly ash particles are coated with layers of amounts of hydration products. However, they are still smooth with different curing times. This finding is different from concrete technology where the precipitation in the pozzolanic reaction is indicated by the etching on fly ash surface (Fraay et al., 1989; Berry et al., 1994; Xu and Sarker, 1994; and Chindapasirt et al., 2005). This is because the input of cement in concrete is high enough to produce a relatively high amount of Ca(OH)$_2$ to be consumed for pozzolanic reaction. Its water to binder ratio (W/B) is generally about 0.2-0.5, providing strength higher than 30 MPa (30,000 kPa) at 28 days of curing, whereas for ground improvement, the W/B is much lower. From this observation, it is thus possible to conclude that the pozzolanic reaction is minimal for strength development in the blended cement stabilized clay.

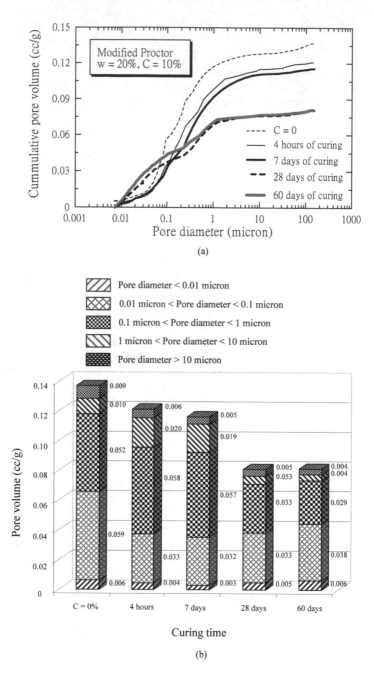

Fig. 9. Pore size distribution of the 10% cement samples compacted at different curing times under modified Proctor energy after 7 days of curing (Horpibulsuk et al., 2010b)

(a) 3% cement   (b) 7% cement

(c) 10% cement

Active Zone

Fig. 10. SEM photos of the cemented samples compacted at different cement contents under modified Proctor energy after 7 days of curing (Horpibulsuk et al., 2010b)

(d) 15% cement   (e) 20% cement

(f) 25% cement   (g) 30% cement

Inert Zone

(h) 35% cement   (i) 40% cement

(j) 45% cement

Decline Zone

Fig. 10. *(Continued)*

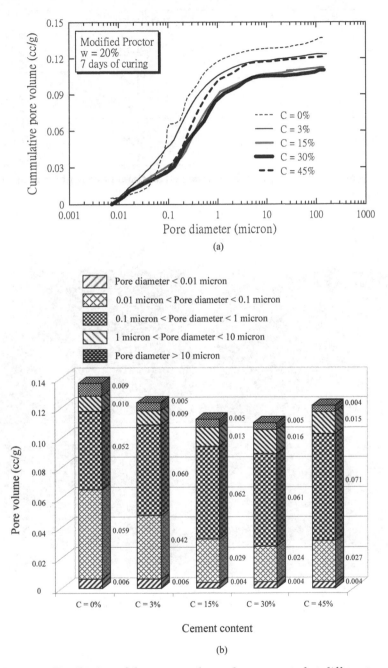

Fig. 11. Pore size distribution of the cemented samples compacted at different water contents and under modified Proctor energy after 7 days of curing time (Horpibulsuk et al., 2010b)

Improvement zones	Cement (%)	Weight loss (%)	Ca(OH)$_2$ (%)
Active	3	1.34	5.51
	7	1.50	6.17
	11	1.60	6.58
Inert	15	1.62	6.66
	20	1.65	6.78
	30	1.68	6.90
Deterioration	35	1.54	6.33
	40	1.48	6.08
	45	1.37	5.63

Table 3. Ca(OH)$_2$ of the cemented samples compacted at different cement contents under modified Proctor energy after 7 days of curing.

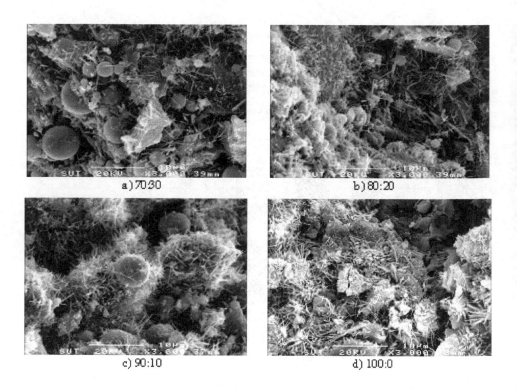

a) 70:30

b) 80:20

c) 90:10

d) 100:0

Fig. 12. SEM photos of the blended cement stabilized clay at different replacement ratios after 28 days of curing (Horpibulsuk et al., 2009)

a) 70:30

b) 80:20

c) 90:10

d) 100:0

Fig. 13. SEM photos of the blended cement stabilized clay at different replacement ratios after 60 days of curing (Horpibulsuk et al., 2010b)

Figure 14 shows the pore size distribution of the CFA blended cement stabilized clay at different curing times and replacement ratios. The pore size distribution for all replacement ratios is almost identical since the grain size distribution and D_{50} of PC and CFA are practically the same. This implies that the strength of blended cement stabilized clay is not directly dependent upon only pore size distribution. However, it might control permeability and durability. With time, the total pore and large pore (>0.1 micron) volumes decrease while the small pore (<0.1 micron) volume increases. This is due to the growth of cementitious products filling up pores.

Table 4 shows $Ca(OH)_2$ of the blended cement stabilized clay at $w = 1.2OWC$ for different curing times. For a particular water content and curing time, $Ca(OH)_2$ for the CFA blended cement stabilized clay decreases with replacement ratio only when the replacement ratio is in excess of a certain value. This finding is different from concrete technology in which $Ca(OH)_2$ decreases significantly with the increase in fineness and replacement ratio (Berry et al., 1989; Sybertz and Wiens, 1991; and Harris et al., 1987; and Chindapasirt et al., 2005 and 2006; and others) due to pozzolanic reaction. The highest $Ca(OH)_2$ is at 10% replacement ratio (C:F = 90:10) for all curing times. For replacement ratios higher than 10%, $Ca(OH)_2$ decreases with replacement ratio. $Ca(OH)_2$ at 20% replacement ratio is almost the same as that at 0% replacement ratio. This finding is associated with the strength test results that the 10% replacement ratio gives the highest strength and the strengths for 0% and 20% replacement ratios are practically the same for all curing times. It is thus concluded that cementitious products mainly control the strength development. In other words, the strengths of the blended cement stabilized clay having different mixing condition (binder content, replacement ratios, and curing time) could be identical as long as cementitious products are the same.

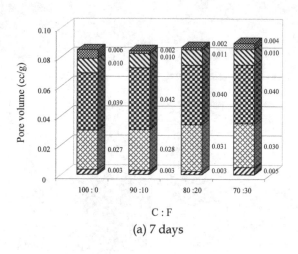

(a) 7 days

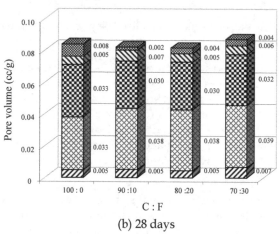

(b) 28 days

	Pore diameter < 0.01 micron
	0.01 micron < Pore diameter < 0.1 micron
	0.1 micron < Pore diameter < 1 micron
	1 micron < Pore diameter < 10 micron
	Pore diameter > 10 micron

Fig. 14. Pore size distribution of the blended cement stabilized clay at different replacement ratios and curing times (Horpibulsuk et al., 2010b)

Curing time (days)	Replacement ratio C : F	Fly ash	Ca(OH)$_2$ (%)		
			Test (Combined effect)	Hydration	Induced (dispersion effect)
7	100:0	-	6.67	6.67	0.00
	90:10	CFA	6.97	6.00	0.97
	80:20	CFA	6.79	5.34	1.45
	70:30	CFA	6.39	4.67	1.72
28	100:0	-	6.79	6.79	0.00
	90:10	CFA	6.96	6.11	0.85
	80:20	CFA	6.81	5.43	1.38
	70:30	CFA	6.57	4.75	1.82
60	100:0	-	6.82	6.82	0.00
	90:10	CFA	7.16	6.14	1.02
	80:20	CFA	6.92	5.46	1.46
	70:30	CFA	6.68	4.77	1.91
90	100:0	-	7.07	7.07	0.00
	90:10	CFA	7.28	6.36	0.91
	80:20	CFA	6.94	5.66	1.28
	70:30	CFA	6.67	4.95	1.72
120	100:0	-	7.08	7.08	0.00
	90:10	CFA	7.29	6.37	0.92
	80:20	CFA	6.96	5.66	1.30
	70:30	CFA	6.70	4.96	1.74

Table 4. Ca(OH)$_2$ of the blended cement stabilized clay at different replacement ratios and curing times.

From SEM and MIP observation, it is notable that the small pore (<0.1 micron) volumes of the blended cement stabilized clay are higher than those of the cement stabilized clay. This implies that a number of large clay-cement clusters possessing large pore space reduce when fly ashes are utilized. In other words, the fly ashes disperse large clay-cement clusters into small clusters, resulting in the increase in small pore volume. The higher the replacement ratio, the better the dispersion. Consequently, the reactive surfaces increase, resulting in the increase in cementitious products as illustrated by dispersion induced $Ca(OH)_2$ (*vide* Table 4). It is the difference in $Ca(OH)_2$ of the blended cement stabilized clay due to the combined effect (hydration and dispersion) and due to hydration. $Ca(OH)_2$ due to combined effect is directly obtained from TG test on the blended cement stabilized sample. $Ca(OH)_2$ due to hydration is also obtained from TG test on the cement stabilized sample having the same cement content as the blended cement stabilized sample. For simplicity, $Ca(OH)_2$ due to hydration at any cement content can be estimated from known $Ca(OH)_2$ of cement stabilized clay at a specific cement content by assuming that the change in the cementitious products is directly proportional to the input of cement (Sinsiri et al., 2006). Thus, $Ca(OH)_2$ due to hydration (H) for any replacement ratio at a particular curing time is approximated in the form.

$$H = T \times (1 - F / 100) \tag{3}$$

where T is known $Ca(OH)_2$ of the cement stabilized clay (0% replacement ratio) obtained from TG test, and F is the replacement ratio expressed in percentage. Sinsiri et al. (2006) have shown that $Ca(OH)_2$ of the cement paste with fly ash is always lower than $Ca(OH)_2$ of the cement paste without fly ash, resulted from $Ca(OH)_2$ consumption for pozzolanic reaction. The same is not for the blended cement stabilized clay. It is found that $Ca(OH)_2$ due to combined effect is higher than that due to hydration for all replacement ratios and curing times. The dispersion induced $Ca(OH)_2$ increases with the replacement ratio for all curing times.

4. Conclusions

This chapter presents the role of curing time, cement content and fly ash content on the strength and microstructure development in the cement stabilized clay. The following conclusions can be advanced:

1. The strength development with cement content for a specific water content is classified into three zones: active, inert and deterioration. In the active zone, the volume of pores smaller than 0.1 micron significantly decreases with the addition of cement because of the increase in cementitious products. In the inert zone, both pore size distribution and cementitious products change insignificantly with increasing cement; thus, there is a slight change in strength. In the deterioration zone, the water is not adequate for hydration because of the excess of cement input. Consequently, as cement content increases, the cementitious products and strength decreases.

2. The flocculation of clay particles due to the cation exchange process is controlled by cement content, regardless of fly ash content. It results in the increase in dry unit weight with insignificant change in liquid limit. Hence, *OWC*s of stabilized and unstabilized silty clay (low swelling clay) are practically the same.

3. The surfaces of fly ash in the blended cement stabilized clay are still smooth for different curing times and fineness, suggesting that pozzolanic reaction is minimal. Fly ash is considered as a dispersing material in the blended cement stabilized clay. This is

different from the application of fly ash as a pozzolanic material in concrete structure in which $Ca(OH)_2$ from hydration is much enough to be consumed for pozzolanic reaction.

4. From the microstructural investigation, it is concluded that the role of fly ash as a non-interacting material is to disperse the cement-clay clusters with large pore space into smaller clusters with smaller pore space. The dispersing effect by fly ash increases the reactive surfaces, and hence the increase in degree of hydration as clearly illustrated by the increase in the induced $Ca(OH)_2$ with replacement ratio and fineness.

5. The increase in cementitious products with time is observed from the scanning electron microscope, mercury intrusion porosimetry and thermal gravity test. With time, the large pore (>0.1 micron) and total pore volumes decrease while the small pore (<0.1 micron) volumes increase. This shows the growth of the cementitious products filling up the large pores.

5. Acknowledgment

This work was a part of the author's researches conducted in the Suranaree University of Technology. The authors would like to acknowledge the financial support provided by the Higher Education Research Promotion and National Research University Project of Thailand, Office of Higher Education Commission, the Thailand Research Fund (TRF), and the Suranaree University of Technology. The author is indebted to Dr. Theerawat Sinsiri, School of Civil Engineering, Suranaree University of Technology for his technical advice in cement and concrete technology. The author is grateful to Mr. Yutthana Raksachon, ex-master's student for his assistance.

6. References

Berry, E.E., Hemmings, R.T., Langley, W.S. & Carette, G.G. (1989). Beneficiated Fly Ash: Hydration, Microstructure, and Strength Development in Portland Cement Systems, in: V.M. Malhotra (Ed.), Third CANMET/ACI, *Conference on Fly Ash, Silica Fume, Slag, and Natural Pozzolans in Concrete* (SP-114), Detroit, pp.241-273.

Chindaprasirt, P., Jaturapitakkul, C. & Sinsiri, T. (2005). Effect of Fly Ash Fineness on Compressive Strength and Pore Size of Blended Cement Plates. *Cement and Concrete Composites*, Vol.27, pp.425-258.

Clough, G.W., Sitar, N., Bachus, R.C. & Rad, N.S. (1981). Cemented Sands under Static Loading. *Journal of Geotechnical Engineering Division*, ASCE, Vol.107, No.GT6, pp.799-817.

El-Jazairi, B. & Illston, J.M. (1977). A Simultaneous Semi-Isothermal Method of Thermogravimetry and Derivative Thermogravimetry, and Its Application to Cement Plates. *Cement and Concrete Research*, Vol.7, pp.247-258.

El-Jazairi, B. & Illston, J.M. (1980). The Hydration of Cement Plate Using the Semi-Isothermal Method of Thermogravimetry. *Cement and Concrete Research*, Vol.10, pp.361-366.

Fraay, A.L.A, Bijen, J.M. & de Haan, Y.M. (1989). The Reaction of Fly Ash in Concrete: A Critical Examination. *Cement and Concrete Research*, Vol.19, pp.235-246.

Gens, A. & Nova, R. (1993). Conceptual Bases for Constitutive Model for Bonded Soil and Weak Rocks, *Geotechnical Engineering of Hard Soil-Soft Rocks*, Balkema.

Kamon, M. & Bergado, D.T. (1992). Ground Improvement Techniques, *Proceedings of 9th Asian Regional Conference on Soil Mechanics and Foundation Engineering*, pp.526-546.

Kasama, K., Ochiai, H. & Yasufuku, N. (2000). On the Stress-Strain Behaviour of Lightly Cemented Clay Based on an Extended Critical State Concept. *Soils and Foundations*, Vol.40, No.5, pp.37-47.

Harris, H.A., Thompson, J.L. & Murphy, T.E. (1987). Factor Affecting the Reactivity of Fly Ash from Western Coals. *Cement and Concrete Aggregate*, Vol.9, pp.34-37.

Horpibulsuk, S. & Miura, N. (2001). A new approach for studying behavior of cement stabilized clays, *Proceedings of 15th international conference on soil mechanics and geotechnical engineering* (ISSMGE), pp. 1759-1762, Istanbul, Turkey.

Horpibulsuk, S., Miura, N. & Nagaraj, T.S. (2003). Assessment of Strength Development in Cement-Admixed High Water Content Clays With Abrams' Law As a Basis. *Geotechnique*, Vol.53, No.4, pp.439-444.

Horpibulsuk, S., Bergado, D.T. & Lorenzo, G.A. (2004a) Compressibility of Cement Admixed Clays at High Water Content. *Geotechnique*, Vol.54, No.2, pp.151-154.

Horpibulsuk, S., Miura, N. & Bergado, D.T. (2004b). Undrained Shear Behavior of Cement Admixed Clay at High Water Content. *Journal of Geotechnical and Geoenvironmental Engineering*, ASCE, Vol.30, No.10, pp.1096-1105.

Horpibulsuk, S., Miura, N., Koga, H. & Nagaraj, T.S. (2004c). Analysis of Strength Development in Deep Mixing – A Field Study. *Ground Improvement Journal*, Vol.8, No.2, pp.59-68.

Horpibulsuk, S., Miura, N. & Nagaraj T.S. (2005). Clay-Water/Cement Ratio Identity of Cement Admixed Soft Clay. *Journal of Geotechnical and Geoenvironmental Engineering*, ASCE, Vol.131, No.2, pp.187-192.

Horpibulsuk, S., Katkan, W., Sirilerdwattana, W. & Rachan, R. (2006). Strength Development in Cement Stabilized Low Plasticity and Coarse Grained Soils : Laboratory and Field Study. *Soils and Foundations*, Vol.46, No.3, pp.351-366.

Horpibulsuk, S., Katkan, W. & Apichatvullop, A. (2008). An Approach for Assessment of Compaction Curves of Fine-Grained Soils at Various Energies Using a One Point Test. *Soils and Foundations*, Vol.48, No.1, pp.115-125.

Horpibulsuk, S., Katkan, W. & Naramitkornburee, A. (2009). Modified Ohio's Curves: A Rapid Estimation of Compaction Curves for Coarse- and Fine-Grained Soils. *Geotechnical Testing Journal*, ASTM, Vol.32, No.1, pp.64-75.

Horpibulsuk, S., Liu, M.D., Liyanapathirana, D.S. & Suebsuk, J. (2010a). Behavior of Cemented Clay Simulated via the Theoretical Framework of the Structured Cam Clay Model. *Computers and Geotechnics*, Vol.37, pp.1-9.

Horpibulsuk, S., Rachan, R., Chinkulkijniwat, A., Raksachon, Y., and Suddeepong, A. (2010b). Analysis of Strength Development in Cement-Stabilized Silty Clay based on Microstructural Considerations. *Construction and Building Materials*, Vol.24, pp.2011-2021.

Horpibulsuk, S., Rachan, R. & Suddeepong, A. (2011a). Assessment of Strength Development in Blended Cement Admixed Bangkok Clay. *Construction and Building Materials*, Vol.25, No.4, pp.1521-1531.

Horpibulsuk, S., Rachan, R., Suddeepong, A. & Chinkulkijniwat, A. (2011b). Strength Development in Cement Admixed Bangkok Clay: Laboratory and Field Investigations. *Soils and Foundations*, Vol.51, No.2, pp.239-251.

JEOL JSM-6400 (1989). Scanning Microscope Operations Manual, Model SM-6400 LOT No. SM150095, Copyright 1989, JEOL Ltd.

Midgley, H.G. (1979). The Determination of Calcium Hydroxide in Set Portland Cements. *Cement and Concrete Research*, Vol.9, pp.77-82.

Mitchell, J.K. (1993). *Fundamentals of Soil Behavior*. John Wiley&Sons, Inc., New York.

Miura, N., Yamadera, A. & Hino, T. (1999). Consideration on Compression Properties of Marine Clay Based on the Pore Size Distribution Measurement. *Journal of Geotechnical Engineering*, JSCE.

Miura, N., Horpibulsuk, S. & Nagaraj, T.S. (2001). Engineering Behavior of Cement Stabilized Clay at High Water Content. *Soils and Foundations*, Vol.41, No.5, pp.33-45.

Nishida, K., Koga, Y. & Miura, N. (1996). Energy Consideration of the Dry Jet Mixing method, *Proceedings of 2nd International Conference on Ground Improvement Geosystems*, IS-Tokyo '96, Vol.1, pp.643-748.

Sinsiri, T., Jaturapitakkul, C. & Chindaprasirt, P. (2006). Influence of Fly Ash Fineness on Calcium Hydroxide in Blended Cement Paste, *Proceedings of Technology and Innovation for Sustainable Development Conference (TISD2006)*, Khon Kaen University, Thailand.

Suebsuk, J., Horpibulsuk, S. & Liu, M.D. (2010). Modified Structured Cam Clay: A Constitutive Model for Destructured, Naturally Structured and Artificially Structured Clays. *Computers and Geotechnics*, Vol.37, pp.956-968.

Suebsuk, J., Horpibulsuk, S. & Liu, M.D. (2011). A Critical State Model for Overconsolidated Structured Clays. *Computers and Geotechnics*, Vol.38, pp.648-658.

Sybert, F. & Wiens, U. (1991). Effect of fly ash fineness on hydration characteristics and strength development, *Proceedings of International Conference on Blended Cement in Construction*, pp.152-165, University of Sheffield, UK.

Wang, K.S., Lin, K.L., Lee, T.Y. & Tzeng, B.Y. (2004). The Hydration Characteristics When C_2S Is Present in MSWI Fly Ash Slag. *Cement and Concrete Research*, Vol.26, pp.323-330.

Washburn, E.W. (1921). Note on Method of Determining the Distribution of Pore Size in Porous Material, *Proceedings of the National Academy of Science*, USA. Vol.7, pp.115-116.

Xu, A. & Sarker, S.L. (1994). Microstructure Development in High-Volume Fly Ash Cement System. *Journal of Material in Civil Engineering*, ASCE, Vol.6, pp.117-136.

Yamadera, A. (1999). *Microstructural Study of Geotechnical Characteristics of Marine Clays*. Ph.D. Dissertation, Saga University, Japan.

Yin, J.H. & Lai, C.K. (1998). Strength and Stiffness of Hong Kong Marine Deposit Mixed With Cement. *Geotechnical Engineering Journal*. Vol.29, No.1, pp.29-44.

6

Evolution of Phases in a Recycled Al-Si Cast Alloy During Solution Treatment

Eva Tillová, Mária Chalupová and Lenka Hurtalová

University of Žilina,
Slovak Republic

1. Introduction

Aluminium has been acquiring increasing significance for the past few decades due to its excellent properties and diversified range of applications. Aluminium has been recognized as one of the best candidate materials for various applications by different sectors such as automotive, construction, aerospace, etc. The increasing demand for aluminium-based products and further globalization of the aluminium industry have contributed significantly to the higher consumption of aluminium scrap for re-production of aluminium alloys (Mahfoud et al., 2010).

Secondary aluminium alloys are made out of aluminium scrap and workable aluminium garbage by recycling. Production of aluminium alloys belong to heavy source fouling of life environs. Care of environment in industry of aluminium connects with the decreasing consumptions resource as energy, materials, waters and soil, with increase recycling and extension life of products. More than half aluminium on the present produce in European Union comes from recycled raw material. By primary aluminium production we need a lot of energy and constraints decision mining of bauxite so European Union has big interest of share recycling aluminium, and therefore increase interest about secondary aluminium alloys and cast stock from them (Sencakova & Vircikova, 2007).

The increase in recycled metal becoming available is a positive trend, as secondary aluminium produced from recycled metal requires only about 2.8 kWh/kg of metal produced while primary aluminium production requires about 45 kWh/kg produced. It is to the aluminium industry's advantage to maximize the amount of recycled metal, for both the energy-savings and the reduction of dependence upon overseas sources. The remelting of recycled metal saves almost 95 % of the energy needed to produce prime aluminium from ore, and, thus, triggers associated reductions in pollution and greenhouse emissions from mining, ore refining, and melting. Increasing the use of recycled metal is also quite important from an ecological standpoint, since producing aluminium by recycling creates only about 5 % as much CO_2 as by primary production (Das, 2006; Das & Gren, 2010).

Today, a large amount of new aluminium products are made by recycled (secondary) alloys. This represents a growing "energy bank" of aluminium available for recycling at the end of components' lives, and thus recycling has become a major issue. The future growth offers an opportunity for new recycling technologies and practices to maximize scrap quality; improve efficiency and reduce cost.

Aluminium-silicon (Al-Si) cast alloys are fast becoming the most universal and popular commercial materials, comprising 85 % to 90 % of the aluminium cast parts produced for the automotive industry, due to their high strength-to-weight ratio, excellent castability, high corrosion resistant and chemical stability, good mechanical properties, machinability and wear resistance. Mg or Cu addition makes Al-Si alloy heat treatable.

The alloys of the Al-Si-Cu system have become increasingly important in recent years, mainly in the automotive industry that uses recycled (secondary) aluminium in the form of various motor mounts, pistons, cylinder heads, heat exchangers, air conditioners, transmissions housings, wheels, fenders and so on due to their high strength at room and high temperature (Rios & Caram, 2003; Li et al., 2004; Michna et al., 2007). The increased use of these recycled alloys demands a better understanding of its response to mechanical properties.

The quality of recycled Al-Si casting alloys is considered to be a key factor in selecting an alloy casting for a particular engineering application. Based on the Al-Si system, the main alloying elements are copper (Cu) or magnesium (Mg) and certain amount of iron (Fe), manganese (Mn) and more, that are present either accidentally, or they are added deliberately to provide special material properties. These elements partly go into solid solution in the matrix and partly form intermetallic particles during solidification. The size, volume and morphology of intermetallic phases are functions of chemistry, solidification conditions and heat treatment (Li, 1996; Paray & Gruzleski, 1994; Tillova & Panuskova, 2007, 2008).

Copper substantially improves strength and hardness in the as-cast and heat-treated conditions. Alloys containing 4 % to 6 % Cu respond most strongly to thermal treatment. Copper generally reduces resistance to general corrosion and, in specific compositions and material conditions, stress corrosion susceptibility. Additions of copper also reduce hot tear resistance and decrease castability. Magnesium is the basis for strength and hardness development in heat-treated Al-Si alloys too and is commonly used in more complex Al-Si alloys containing copper, nickel, and other elements for the same purpose.

Iron considers the principal impurity and detrimental alloying element for Al-Si-Cu alloys. Iron improves hot tear resistance and decreases the tendency for die sticking or soldering in die casting. Increases in iron content are, however, accompanied by substantially decreased ductility. Iron reacts to form a myriad of insoluble phases in aluminium alloy melts, the most common of which are Al_3Fe, Al_6FeMn, and α-Al_5FeSi. These essentially insoluble phases are responsible for improvements in strength, especially at elevated temperature. As the fraction of insoluble phase increases with increased iron content, casting considerations such as flowability and feeding characteristics are adversely affected. Iron also lead to the formation of excessive shrinkage porosity defects in castings (Warmuzek, 2004a; Taylor, 2004; Shabestari, 2004; Caceres et al., 2003; Wang et al. 2001; Tillova & Chalupova, 2010).

It is clear that the morphology of Fe-rich intermetallic phases influences harmfully also fatigue properties (Taylor, 2004; Tillova & Chalupova, 2010). It is recognized that recycled Al-Si-Cu alloys are not likely to be suitable for fracture-critical components, where higher levels of Fe and Si have been shown to degrade fracture resistance. However the likelihood exists that they may perform quite satisfactorily in applications such as those listed where service life is determined by other factors (Taylor, 2004).

2. Experimental material and methodology

As an experimental material recycled (secondary) hypoeutectic AlSi9Cu3 alloy, in the form of 12.5 kg ingots, was used. The alloy was molten into the sand form (sand casting). The melting temperature was maintained at 760 °C ± 5 °C. Molten metal was before casting purified with salt AlCu4B6. The melt was not modified or grain refined. The chemical analysis of AlSi9Cu3 cast alloy was carried out using arc spark spectroscopy. The chemical composition is given in the table 1.

Si	Cu	Mn	Fe	Mg	Ni	Pb	Zn	Ti	Al
10.7	2.4	0.22	< 0.8	0.47	0.08	0.11	1.1	0.03	rest

Table 1. Chemical composition of the alloy (wt. %)

AlSi9Cu3 cast alloy has lower corrosion resistance and is suitable for high temperature applications (dynamic exposed casts, where are not so big requirements on mechanical properties) - it means to max. 250 °C. Experimental samples (standard tensile test specimens) were given a T4 heat treatment - solution treatment for 2, 4, 8, 16 or 32 hours at three temperatures (505 °C, 515 °C and 525 °C); water quenching at 40 °C and natural aging for 24 hours at room temperature. After heat treatment samples were subjected to mechanical test. For as cast state, each solution temperature and each aging time, a minimum of five specimens were tested.

Metallographic samples were prepared from selected tensile specimens (after testing) and the microstructures were examined by optical (Neophot 32) and scanning electron microscopy. Samples were prepared by standards metallographic procedures (mounting in bakelite, wet ground, DP polished with 3 μm diamond pastes, finally polished with commercial fine silica slurry (STRUERS OP-U) and etched by Dix-Keller. For setting of Fe-rich intermetallic phases was used etching by H_2SO_4. For setting of Cu-rich intermetallic phases was used etching by HNO_3.

Some samples were also deep-etched for 30 s in HCl solution in order to reveal the three-dimensional morphology of the eutectic silicon and intermetallic phases (Tillova & Chalupova, 2001, 2009). The specimen preparation procedure for deep-etching consists of dissolving the aluminium matrix in a reagent that will not attack the eutectic components or intermetallic phases. The residuals of the etching products should be removed by intensive rinsing in alcohol. The preliminary preparation of the specimen is not necessary, but removing the superficial deformed or contaminated layer can shorten the process. To determine the chemical composition of the intermetallic phases was employed scanning electron microscopy (SEM) TESCAN VEGA LMU with EDX analyser BRUKER QUANTAX.

Quantitative metallography (Skocovsky & Vasko, 2007; Vasko & Belan, 2007; Belan, 2008; Vasko, 2008; Martinkovic, 2010) was carried out on an Image Analyzer NIS - Elements 3.0 to quantify phase's changes during heat treatment. A minimum of 20 pictures at 500 x magnification of the polish per specimen were taken.

Hardness measurement was preformed by a Brinell hardness tester with a load of 62.5 kp, 2.5 mm diameter ball and a dwell time of 15 s. The Brinell hardness value at each state was obtained by an average of at least six measurements. The phases Vickers microhardness was measured using a MHT-1 microhardness tester under a 1g load for 10 s (HV 0.01). Twenty measurements were taken per sample and the median microhardness was determined.

3. Results and discussion

3.1 Microstructure of recycled AlSi9Cu3 cast alloy

Controlling the microstructure during solidification is, therefore, very important. The Al-Si eutectic and intermetallic phases form during the final stage of the solidification. How the eutectic nucleates and grows have been shown to have an effect on the formation of defects such as porosity and microporosity too. The defects, the morphology of eutectic and the morphology of intermetallic phases have an important effect on the ultimate mechanical properties of the casting.

As recycling of aluminium alloys becomes more common, sludge will be a problem of increasing importance due to the concentration of Fe, Mn, Cr and Si in the scrap cycle. During the industrial processing of the Al-Si alloys, these elements go into solid solution but they also form different intermetallic phases. The formation of these phases should correspond to successive reaction during solidification - table 2 (Krupiński et al., 2011; Maniara et al., 2007; Mrówka-Nowotnik & Sieniawski, 2011; Dobrzański et al., 2007, Tillova & Chalupova, 2009). Thus, control of these phases e. g. quantitative analysis (Vasko & Belan, 2007; Martinkovic, 2010) is of considerable technological importance. Typical structures of the recycled as-cast AlSi9Cu3 alloys are shown in Fig. 1. The microstructure consists of dendrites α-phase (1), eutectic (mixture of α-matrix and spherical Si-phases - 2) and variously type's intermetallic Fe- and Cu-rich phases (3 and 4).

Reactions	Temperature, °C
α - dendritic network	609
Liq. → α - phase + $Al_{15}Mn_3Si_2$ + Al_5FeSi	590
Liq. → α - phase + Si + Al_5FeSi	575
Liq. → α - phase + Al_2Cu + Al_5FeSi + Si	525
Liq. → α - phase + Al_2Cu + Si + $Al_5Mg_8Si_6Cu_2$	507

Table 2. Reactions occurring during the solidification of AlSi9Cu3 alloys

The α-matrix precipitates from the liquid as the primary phase in the form of dendrites and is nominally comprised of Al and Si. Experimental material was not modified and so eutectic Si particles are in a form of platelets (Fig. 2a), which on scratch pattern are in a form of needles – Fig. 2b (Skocovsky et al., 2009; Tillova & Chalupova, 2001; 2009).

Iron is one of the most critical alloying elements, because Fe is the most common and usually detrimental impurity in cast Al-Si alloys. Iron impurities can either come from the use of steel tools or scrap materials or be acquired during subsequent melting, remelting and casting, e.g. by contamination from the melting pot etc.

A number of Fe-rich intermetallic phases, including α (Al$_8$Fe$_2$Si or Al$_{15}$(FeMn)$_3$Si$_2$), β (Al$_5$FeSi), π (Al$_8$Mg$_3$FeSi$_6$), and δ (Al$_4$FeSi$_2$), have been identified in Al-Si cast alloys (Samuel et al., 1996; Taylor, 2004; Seifeddine, 2007; Seifeddine et al. 2008; Moustafa, 2009; Fang et al., 2007; Lu & Dahle, 2005).

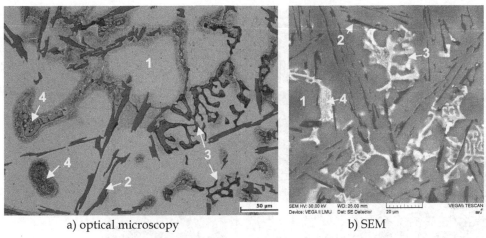

a) optical microscopy   b) SEM

Fig. 1. Microstructure of recycled AlSi9Cu3 cast alloy (1 – α-phase, 2 – eutectic silicon, 3 – Fe-rich phases, 4 – Cu-rich phases), etch. Dix-Keller

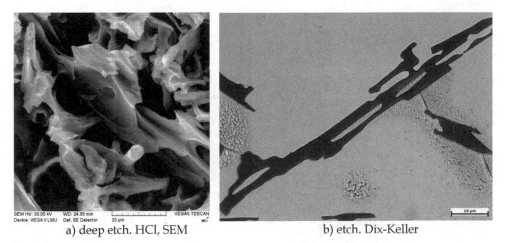

a) deep etch. HCl, SEM   b) etch. Dix-Keller

Fig. 2. Morphology of eutectic silicon

In experimental AlSi9Cu3 alloy was observed the two main types of Fe-rich intermetallic phases, Al$_5$FeSi with monoclinic crystal structure (know as beta- or β-phase) and Al$_{15}$(FeMn)$_3$Si$_2$ (know as alpha- or α-phase) with cubic crystal structure. The first phase (Al$_5$FeSi) precipitates in the interdendritic and intergranular regions as platelets (appearing as needles in the metallographic microscope - Fig. 3). Long and brittle Al$_5$FeSi platelets (more than 500 µm) can adversely affect mechanical properties, especially ductility, and also lead to the formation of excessive shrinkage porosity defects in castings (Caceres et al., 2003). Platelets are effective pore nucleation sites. It was also shown that the Al_5FeSi needles can act as nucleation sites for Cu-rich Al_2Cu phases (Tillova et al., 2010).

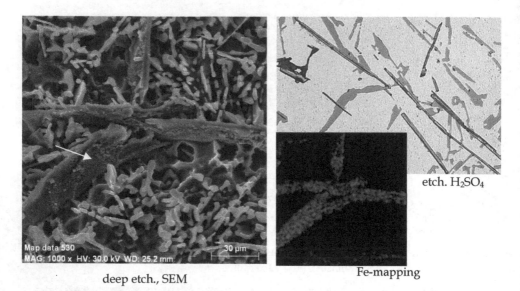

deep etch., SEM   Fe-mapping

Fig. 3. Morphology of Fe-phase Al_5FeSi

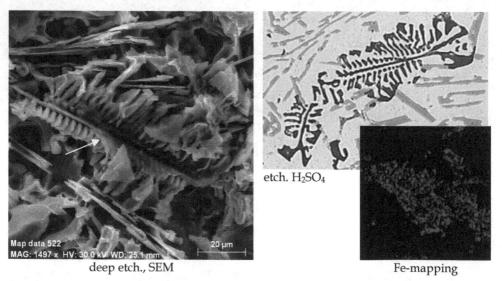

deep etch., SEM   Fe-mapping

Fig. 4. Morphology of Fe-phase $Al_{15}(FeMn)_3Si_2$

The deleterious effect of Al_5FeSi can be reduced by increasing the cooling rate, superheating the molten metal, or by the addition of a suitable "neutralizer" like Mn, Co, Cr, Ni, V, Mo and Be. The most common addition has been manganese. Excess Mn may reduce Al_5FeSi phase and promote formation Fe-rich phases $Al_{15}(FeMn)_3Si_2$ in form „skeleton like" or in form „Chinese script" (Seifeddine et al., 2008; Taylor, 2004) (Fig. 4). This compact morphology "Chinese script" (or skeleton - like) does not initiate cracks in the cast material to the same extent as Al_5FeSi does and phase $Al_{15}(FeMn)_3Si_2$ is considered less harmful to the mechanical properties than β phase (Ma et al., 2008; Kim et al., 2006). The amount of manganese needed to neutralize iron is not well established. A common "rule of thumb" appears to be ratio between iron and manganese concentration of 2:1.

Alloying with Mn and Cr, caution has to be taken in order to avoid the formation of hard complex intermetallic multi-component sludge, $Al_{15}(FeMnCr)_3Si_2$ - phase (Fig. 5). These intermetallic compounds are hard and can adversely affect the overall properties of the casting. The formation of sludge phases is a temperature dependent process in a combination with the concentrations of iron, manganese and chromium independent of the silicon content. If Mg is also present with Si, an alternative called pi-or π-phase can form, $Al_5Si_6Mg_8Fe_2$. $Al_5Si_6Mg_8Fe_2$ has a script-like morphology. The Fe-rich particles can be twice as large as the Si particles, and the cooling rate has a direct impact on the kinetics, quantities and size of Fe-rich intermetallic present in the microstructure.

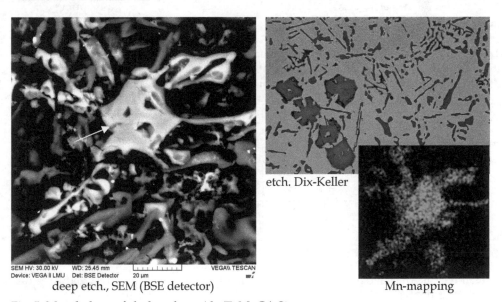

etch. Dix-Keller

SEM HV: 30.00 KV   WD: 25.45 mm   VEGA\\ TESCAN
Device: VEGA II LMU   Det: BSE Detector   20 μm

deep etch., SEM (BSE detector)   Mn-mapping

Fig. 5. Morphology of sludge phase $Al_{15}(FeMnCr)_3Si_2$

Cu is in Al-Si-Cu cast alloys present primarily as phases: Al_2Cu, $Al-Al_2Cu-Si$ or $Al_5Mg_8Cu_2Si_6$ (Rios et al., 2003; Tillova & Chalupova, 2009; Tillova et al.; 2010). The average size of the Cu-phase decreases upon Sr modification. The Al_2Cu phase is often observed to precipitate both in a small blocky shape with microhardness 185 HV 0.01. $Al-Al_2Cu-Si$ phase is observed in very fine multi-phase eutectic-like deposits with microhardness 280 HV 0.01

(Tillova & Chalupova, 2009). In recycled AlSi9Cu3 alloy was analysed two Cu-phases: Al_2Cu and $Al-Al_2Cu-Si$ (Fig. 6).

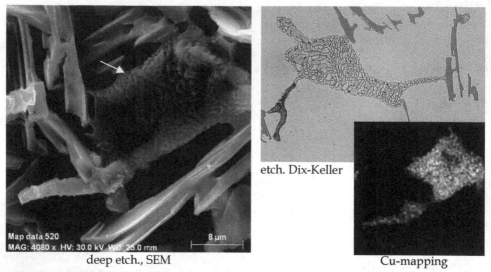

deep etch., SEM   Cu-mapping

Fig. 6. Morphology of Cu-phase - $Al-Al_2Cu-Si$

The microhardness of all observed intermetallic phases was measured in HtW Dresden and the microhardness values are indicated in table 3. It is evident that the eutectic silicon, the Fe-rich phase Al_5FeSi and the multicomponent intermetallic $Al_{15}(FeMn)_3Si_2$ are the hardest.

Intermetallic phases	HV 0.01	Chemical composition, wt. %					
		Al	Mg	Si	Fe	Cu	Mn
$Al_{15}(MnFe)_3Si_2$	483	61	-	10.3	13.4	2.6	13.6
Al_5FeSi	1 475	67.7	-	16.5	15.8	-	-
Al_2Cu	185	53.5	-	-	-	42.2	-
$Al-Al_2Cu-Si$	280	53	4.5	14.8	-	18.5	-
Si	1084	-	-	99.5	-	-	-

Table 3. Microhardness and chemical composition of intermetallic phases

Influence of intermetallic phases to mechanical and fatigue properties of recycled Al-Si cast alloys depends on size, volume and morphology this Fe- and Cu-rich phases.

3.2 Effect of solution treatment on the mechanical properties

Al-Si-Cu cast alloys are usually heat-treated in order to obtain an optimum combination of strength and ductility. Important attribute of a precipitation hardening alloy system is a temperature and time dependent equilibrium solid-solubility characterized by decreasing solubility with decreasing temperature and then followed by solid-state precipitation of second phase atoms on cooling in the solidus region (Abdulwahab, 2008; Michna et al., 2007; ASM Handbook, 1991). Hardening heat treatment involves (Fig. 7):

- Solution heat - treatment - it is necessary to produce a solid solution. Production of a solid solution consists of soaking the aluminium alloy at a temperature sufficiently high and for such a time so as to attain an almost homogeneous solid solution;
- Rapid quenching to retain the maximum concentration of hardening constituent (Al$_2$Cu) in solid solution. By quenching it is necessary to avoid slow cooling. Slow cooling can may the precipitation of phases that may be detrimental to the mechanical properties. For these reasons solid solutions formed during solution heat-treatment are quenched rapidly without interruption to produce a supersaturated solution at room temperature;
- Combination of artificial and over-ageing to obtain the desired mechanical properties in the casting. Generally artificial aging imparts higher strength and hardness values to aluminium alloys without sacrificing other mechanical properties.

The precipitation sequence for Al-Si-Cu alloy is based upon the formation of Al$_2$Cu based precipitates. The sequence is described as: $\alpha_{ss} \rightarrow$ GP zones $\rightarrow \theta' \rightarrow \theta$ (Al$_2$Cu). The sequence begins upon aging when the supersaturated solid solution (α_{ss}) gives way first to small coherent precipitates called GP zones. These particles are invisible in the optical microscope but macroscopically, this change is observed as an increase in the hardness and tensile strength of the alloy. As the process proceeds, the GP zones start to dissolve, and θ' begins to form, which results in a further increase in the hardness and tensile strength in the alloy. Continued aging causes the θ' phase to coarsen and the θ (Al$_2$Cu) precipitate to appear. The θ phase is completely incoherent with the matrix, has a relatively large size, and has a coarse distribution within the aluminium matrix. Macroscopically, this change is observed as an increase in the ductility and a decrease in the hardness and tensile strength of the alloy (Abdulwahab, 2008; Michna et al., 2007; Panuskova et al., 2008).

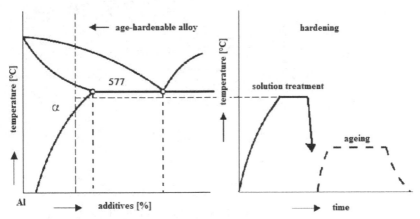

Fig. 7. The schematic diagram of hardening process for Al-Si-Cu cast alloy

Although the morphology, the amount and the distribution of the precipitates during aging process significantly influence the mechanical properties, an appropriate solution treatment is a prerequisite for obtaining desirable aging effect. From this point of view, the solution heat treatment is critical in determining the final microstructure and mechanical properties of the alloys. Thus, it is very important to investigate the effects of solution heat treatment on the alloys, before moving on to aging issues.

Solution treatment performs three roles (Li, 1996; Lasa & Rodriguez-Ibabe, 2004; Paray & Gruzleski, 1994; Moustafa et al., 2003; Sjölander & Seifeddine, 2010):

- homogenization of as-cast structure;
- dissolution of certain intermetallic phases such as Al_2Cu;
- changes the morphology of eutectic Si phase by fragmentation, spheroidization and coarsening, thereby improving mechanical properties, particularly ductility.

For experimental work heat treatment consisted of solution treatment for different temperatures: 505 °C, 515 °C and 525 °C; rapid water quenching (40 °C) and natural ageing (24 hours at room temperature) was used.

Influence of solution treatment on mechanical properties (strength tensile - R_m and Brinell hardness - HBS) is shown in Fig. 8 and Fig. 9.

After solution treatment, tensile strength, ductility and hardness are remarkably improved, compared to the corresponding as-cast condition. Fig. 8 shows the results of tensile strength measurements. The as cast samples have a strength value approximately 204 MPa. After 2 hours the solution treatment, independently of temperature of solution treatment, strength value immediately increases. By increasing the solution holding time from 2 to 4 hours, the tensile strength increased to 273 MPa for 515 °C. With further increase in solution temperature more than 515 °C and solution treatment time more than 4 hours, tensile strength decreases during the whole period as a result of gradual coarsening of eutectic Si, increase of inter particle spacing and dissolution of the Al_2Cu phase (at 525 °C).

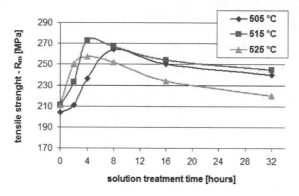

Fig. 8. Influence of solution treatment conditions on tensile strength

Fig. 9 shows the evolution of Brinell hardness value. Results of hardness are comparable with results of tensile strength. The as cast samples have a hardness value approximately 98 HB. After 2 hours the solution treatment, independently from temperature of solution treatment, hardness value immediately increases. The maximum was observed after 4 hours - approximately 124 HBS for 515 °C. However, after 8 hours solution treatment, the HB values are continuously decreasing as results of the coarsening of eutectic silicon, increase of inter particle spacing and dissolution of the Al₂Cu phase. After prolonged solution treatment time up to 16 h at 525 °C, it is clearly that the HB values strongly decrease probably due to melting of the Al-Al₂Cu-Si phase.

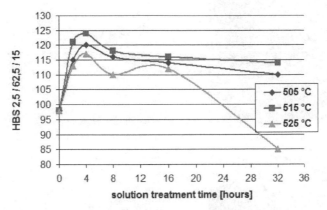

Fig. 9. Influence of solution treatment conditions on Brinell hardness

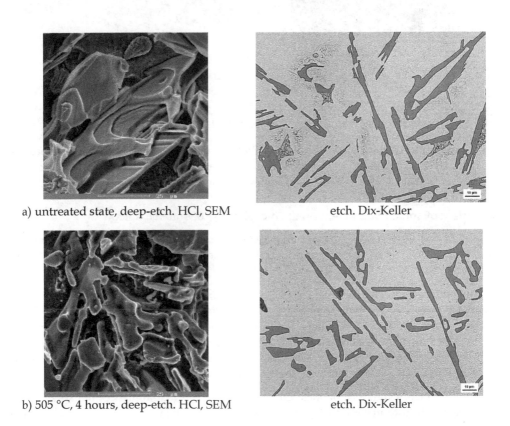

a) untreated state, deep-etch. HCl, SEM   etch. Dix-Keller

b) 505 °C, 4 hours, deep-etch. HCl, SEM   etch. Dix-Keller

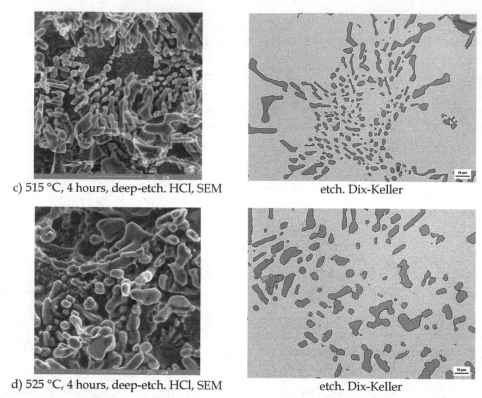

c) 515 °C, 4 hours, deep-etch. HCl, SEM   etch. Dix-Keller

d) 525 °C, 4 hours, deep-etch. HCl, SEM   etch. Dix-Keller

Fig. 10. Effect of solution treatment on morphology of eutectic Si

Obtained results (Fig. 8 and Fig. 9) suggests that to enhance the tensile strength or hardness of this recycled cast alloy by increasing of solution temperature more than 515 °C and by extending the solution time more than 4 hours does not seem suitable.

3.3 Effect of solution treatment on the morphology of eutectic silicon

The mechanical properties of cast component are determined largely by the shape and distribution of Si particles in the matrix. Optimum tensile, impact and fatigue properties are obtained with small, spherical and evenly distributed particles.

It is hypothesized (Paray & Gruzleski, 1994; Li, 1996; Tillova & Chalupova, 2009; Moustafa et al, 2010) that the spheroidisation process of the eutectic silicon throughout heat treatment takes place in two stages: fragmentation or dissolution of the eutectic Si branches and the spheroidisation of the separated branches. Experimental material was not modified or grain refined and so eutectic Si particles without heat treatment (untreated – as cast state) are in a form of platelets (Fig. 10a), which on scratch pattern are in a form of needles.

The solution temperature is the most important parameter that influences the kinetics of Si morphology transformation during the course of solution treatment. The effect of solution treatment on morphology of eutectic Si, for holding time 4 hour, is demonstrated in Figures 10b, 10c and 10d. After solution treatment at the temperature of 505 °C were noted that the platelets were fragmentized into smaller platelets with spherical edges (Fig. 10b) (on scratch pattern round needles). The temperature 505 °C is for Si-spheroidisation low.

The spheroidisation process dominated at 515 °C. Si platelets fragment into smaller segments and these smaller Si particles were spheroidised to rounded shape; see Fig. 10c. By solution treatment 525 °C the spheroidised particles gradually grew larger (coarsening) (Figures 10d).

Quantitative metallography (Skocovsky & Vasko, 2007; Vasko & Belan, 2007; Belan, 2008; Vasko, 2008; Martinkovic, 2010) was carried out on an Image Analyzer NIS-Elements to quantify eutectic Si (average area of eutectic Si particle) by magnification 500 x. Figure 11 shows the average area of eutectic Si particles obtained in solution heat treated samples. This graphic relation is in line with work Paray & Gruzleski, 1994.

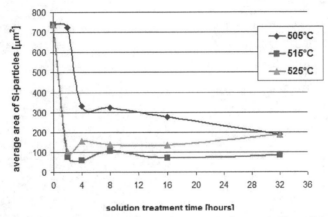

Fig. 11. Influence of solution treatment on average area of eutectic Si particles

Average area of eutectic Si particles decreases with increasing solution temperature and during the whole solution period. During the two hours, the area of Si-particles decreases which indicated that they undergo fragmentation and break into smaller segments.

Minimum value of average eutectic Si particles was observed by temperature 515 °C (approximately 89 μm^2). It's probably context with spheroidisation of eutectic silicon on this temperature. By solution treatment 525°C the spheroidised Si-particles in comparison with temperature 515 °C coarsen. The value of average eutectic Si particles at this temperature was observed from approximately 100 μm^2 (2 hour) till 187 μm^2 (32 hour). Prolonged solution treatment at 515°C and 525°C leads to a significant coarsening of the spheroidised Si particles.

3.4 Effect of solution treatment on the morphology of Fe-rich phases

The influence of iron on mechanical properties of aluminium alloys depends on the type, morphology and quantity of iron in the melt. Nevertheless, the shape of iron phases is more influential than the quantity of those iron compounds.

The evolution of the Fe-rich phases during solution treatment is described for holding time 4 hours in Fig. 12. Al_5FeSi phase is dissolved into very small needles (difficult to observe). The $Al_{15}(FeMn)_3Si_2$ phase was fragmented to smaller skeleton particles. In the untreated state $Al_{15}(FeMn)_3Si_2$ phase has a compact skeleton-like form (Fig. 12a). Solution treatment of this skeleton-like phase by 505°C tends only to fragmentation (Fig. 12b) and by 515°C or 525°C to fragmentation, segmentation and dissolution (Fig. 12c, Fig. 12d).

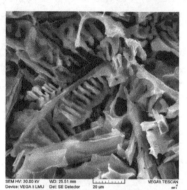

a) untreated state, deep-etch. HCl, SEM

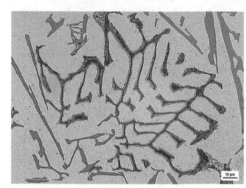

etch. H_2SO_4

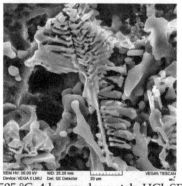

b) 505 °C, 4 hours, deep-etch. HCl, SEM

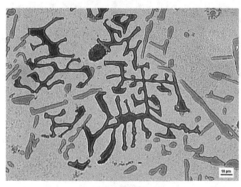

etch. H_2SO_4

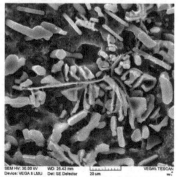

c) 515 °C, 4 hours, deep-etch. HCl, SEM

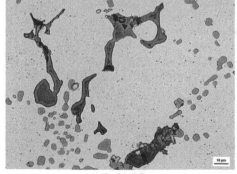

etch. H_2SO_4

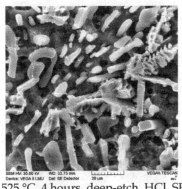
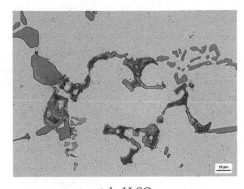

d) 525 °C, 4 hours, deep-etch. HCl, SEM   etch. H₂SO₄

Fig. 12. Effect of solution treatment on morphology of Fe-rich phases

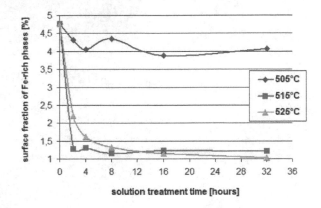

Fig. 13. Influence of solution treatment on surface fraction of Fe-rich phases

Quantitative metallography was carried out on an Image Analyzer NIS-Elements to quantify Fe-phases changes, during solution treatment. It was established that the temperature increase of solution treatment was attended not only by fragmentation of $Al_{15}(FeMn)_3Si_2$ phase, but also by decrease of surface fraction of all Fe-rich phases in AlSi9Cu3 alloy (Fig. 13). For the non-heat treated state the surface fraction of Fe-rich phase was c. 4.8 %, for temperature 515 °C c. 1.6 % and for 525 °C only c. 1.25 %. Solution treatment reduces its surface fraction rather than changes its morphology (Fig. 12 and Fig. 13).

3.5 Effect of solution treatment o the morphology of Cu-rich phases

The Cu-rich phase solidifies as fine ternary eutectic (Al-Al₂Cu-Si) - Fig. 6. Effect of solution treatment on morphology of Al-Al₂Cu-Si is demonstrated on Fig. 14. The changes of morphology of Al-Al₂Cu-Si observed after heat treatment are documented for holding time 4 hours. Al-Al₂Cu-Si phase without heat treatment (untreated state) occurs in form compact oval troops (Figures 14a and 15a).

After solution treatment by temperature 505 °C these phase disintegrated into fine smaller segments and the amount of Al-Al$_2$Cu-Si phase during heat treatment decreases. This phase is gradually dissolved into the surrounding Al-matrix with an increase in solution treatment time (Fig. 14b). By solution treatment by 515 °C is this phase observed in the form coarsened globular particles and these occurs along the black needles, probably Fe-rich Al$_5$FeSi phase (Figures 14c and 15b). By solution treatment 525 °C is this phase documented in the form molten particles with homogenous shape (Fig. 14d).

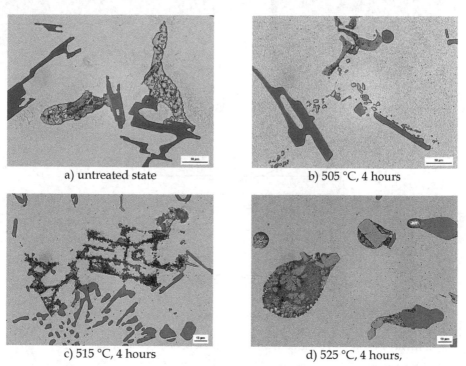

a) untreated state   b) 505 °C, 4 hours

c) 515 °C, 4 hours   d) 525 °C, 4 hours,

Fig. 14. Effect of solution treatment on morphology of Cu-rich phases, etch. HNO$_3$

Compact Al-Al$_2$Cu-Si phase disintegrates to fine separates Al$_2$Cu particles. The amount of these phases was not obvious visible on optical microscope. On SEM microscope we observed these phases in form very small particles for every temperatures of natural aging. By observation we had to use a big extension, because we did not see these elements.

Small precipitates (Al$_2$Cu) incipient by hardening were invisible in the optical microscope and electron microscope so it is necessary observation using TEM microscopy.

3.6 SEM observation of the fracture surface

Topography of fracture surfaces is commonly examined by SEM. The large depth of field is a very important advantage for fractographic investigations. Fracture surfaces of Al-Si-Cu cast alloys can be observed by means of SEM without almost any special preparation; nevertheless, if it is possible, the specimens should be examined immediately after failure because of the very fast superificial oxidation of Al-alloys. In some cases, the specimen should be cleaned mechanically by rinsing in ultrasonic cleaner, chemical reagents, or electrolytes (Michna et al., 2007; Tillova & Chalupova, 2009; Warmuzek, 2004b).

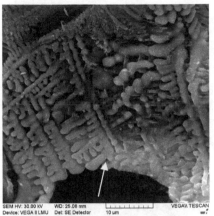

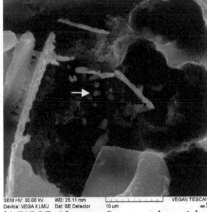

a) untreated state – compact morphology   b) 515 °C, 4 hours - fine round particles

Fig. 15. Morphology of Cu-rich phases after deep etching, etch. HCl, SEM

Fractographs of the specimens in untreated state (as cast state) after impact test are documented in Fig. 16. As the experimental material was not modified and eutectic Si particles are in a form of platelets (Fig. 2), fracture surfaces are mainly composed of ductile fracture with cleavage fracture regions.

Fracture of the α-matrix is transcrystalline ductile with dimples morphology and with plastically transformed walls (Fig. 16a, b). The shape of walls depends on the orientation of Si particles on fracture surface. The brittle eutectic Si and Fe-rich phases (Figures 3-5) are fractured by the transcrystalline cleavage mechanism (Fig. 16c, d, e). Cu-phase (compact ternary eutectic Al-Al$_2$Cu-Si – Fig. 6) is fractured by transcrystalline ductile fracture with the very fine and flat dimples morphology (Fig. 16f). In some cases, to improve the contrast of the matrix/phase interface, detection of backscattered electrons (BSE) in a SEM is a very useful method (Fig. 16c). This method provides another alternative when phase attribution by morphology and/or colour, is not clearly.

Fractographs of the specimens after solution treatment are documented in Fig. 17. By temperature 505 °C of solution treatment were noted that the Si-platelets were fragmentized into smaller platelets with spherical edges (Fig. 10b). Spheroidisation process of eutectic silicon was not observed. The morphology from transcrystalline brittle fracture (cleavage) is mainly visible, but some degree of plastic deformation in the aluminium solid solution (α-matrix) also may be noticed in the form of shallow dimples and plastically transformed walls (Fig. 17a, b).

After solution treatment at the 515 °C eutectic silicon is completely spheroidised (Fig. 10c). Number of brittle Fe-phases decreases (Fig. 12c). Fracture is transcrystalline ductile with fine dimples morphology (Fig. 17c, d). The size of the dimples shows the size of eutectic silicon. Local we can observe little cleavage facets of Fe-rich phases.

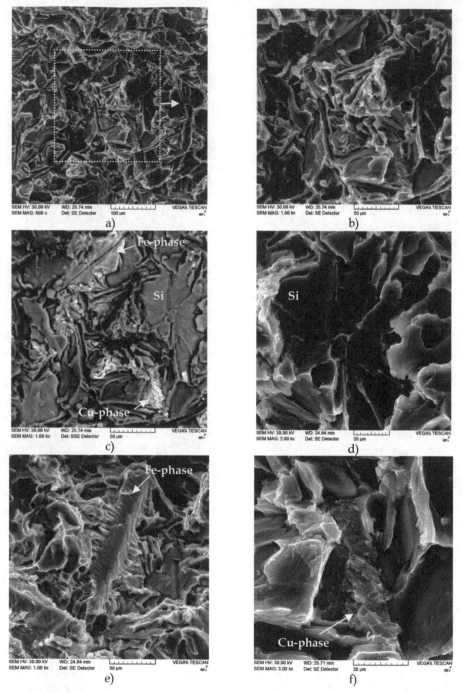

Fig. 16. Fractographs of the impact test specimen – as cast state, SEM

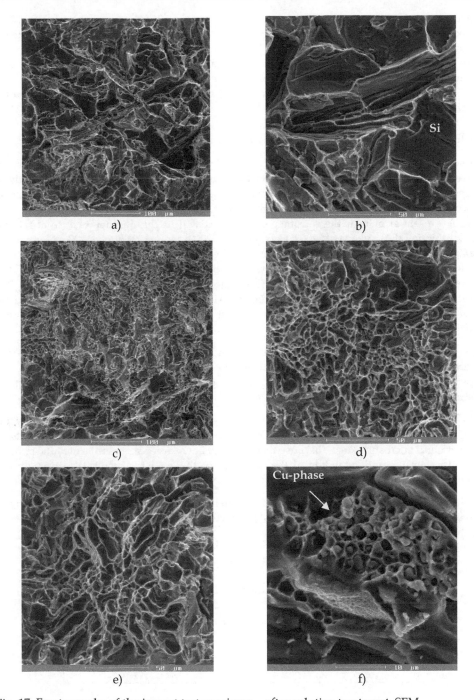

Fig. 17. Fractographs of the impact test specimen – after solution treatment, SEM

Fractograph of the specimen after solution treatment at the 525 °C is documented in Fig. 17e. Eutectic silicon is completely spheroidised too (Fig. 10d), but spheroidised Si-particles gradually grew larger. The fracture mechanism was identified as transcrystalline ductile with dimples morphology accompanied by plastically transformed walls (Fig. 17e). The size of the dimples shows the larger size of eutectic silicon as compared with fractograph Fig. 17d. Figure 17f is an example of a transcrystalline ductile fracture of Cu-rich phase after solution treatment at the 515 °C.

3.7 Influence of solution annealing on fatigue properties

To successfully utilize recycled Al-Si-Cu alloys in critical components, it is necessary to thoroughly understand its fatigue property too. Numerous studies have shown that fatigue property of conventional casting aluminium alloys are very sensitive to casting defects (porosity, microshrinkages and voids) and many studies have shown that, whenever a large pore is present at or near the specimen's surface, it will be the dominant cause of fatigue crack initiation (Bokuvka et al., 2002; Caceres et al., 2003; Moreira & Fuoco, 2006; Novy et al.; 2007). The occurrence of cast defects, together with the morphology of microstructural features, is strongly connected with method of casting too. By sand mould is the concentration of hydrogen in melt, as a result of damp cast surroundings, very high. The solubility of hydrogen during solidification of Al-Si cast alloys rapidly decreases and by slow cooling rates (sand casting) keeps in melt in form of pores and microshrinkages (Michna et al., 2007).

Fe is a common impurity in aluminium alloys that leads to the formation of complex Fe-rich intermetallic phases, and how these phases can adversely affect mechanical properties, especially ductility, and also lead to the formation of excessive shrinkage porosity defects in castings (Taylor, 2004; Tillova & Chalupova, 2009).

It is clear, that the morphology of Fe-rich intermetallic phases influences harmfully on fatigue properties too (Palcek et. al., 2003). Much harmful effect proves the cast defects as porosity and microshrinkages, because these defects have larger size as intermetallic phases. A comprehensive understanding of the influence of these microstructural features on the fatigue damage evolution is needed.

In the end heat treatment is considered as an important factor that affects the fatigue behaviour of casting Al-Si-Cu alloys too (Tillova & Chalupova, 2010).

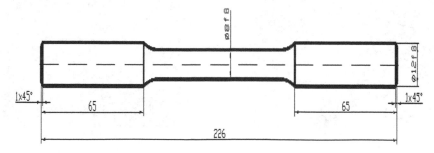

Fig. 18. Fatigue specimen geometry (all dimension in mm)

The fatigue AlSi9Cu3 tests (as-cast, solution heat treated at two temperatures 515 and 525 °C for times 4 hours, then quenched in warm water at 40 °C and natural aged at room temperature for 24 hours) were performed on rotating bending testing machine ROTOFLEX operating at 30 Hz., load ratio R = -1 and at room temperature 20 ± 5 °C on the air. Cylindrical fatigue specimens were produced by lathe-turning and thereafter were heat treated. Geometry of fatigue specimens is given in Fig. 18. The fatigue fracture surfaces of the fatigue - tested samples under different solution heat treatment condition were examined using a scanning electron microscope (SEM) TESCAN VEGA LMU with EDX analyser BRUKER QUANTAX after fatigue test.

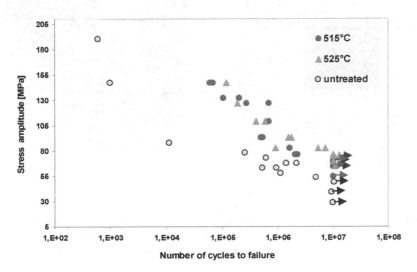

Fig. 19. Effect of solution treatment on fatigue behaviour of AlSi9Cu3 cast alloy

The untreated specimens were tested first to provide a baseline on fatigue life. In this study, the number of cycle, 10^7, is taken as the infinite fatigue life. Thus, the highest applied stress under which a specimen can withstand 10^7 cycles is defined as the fatigue strength of the alloy. The relationship between the maximum stress level (S), and the fatigue life in the form of the number of fatigue cycles (N), (S-N curves) is given in Fig. 19. Comparison on the fatigue properties of specimens with and without heat treatment was made. In heat untreated state has fatigue strength (σ) at 10^7 cycles the lowest value, only σ = 49 MPa. It is evident, that after solution treatment increased fatigue strength at 10^7 cycles. By the conditions of solution treatment 515 °C/4 hours the fatigue strength at 10^7 cycles increases up to value σ = 70 MPa. The solution treatment by 525 °C/4 hours caused the increasing of fatigue strength at 10^7 cycles to value σ = 76 MPa. The growths in fatigue strength at 10^7 cycles with respect to the temperature of solution treatment are 42 and 55 % respectively.

Fatigue fracture surfaces were examined in the SEM in order to find the features responsible for crack initiation. Typical fractographic surfaces are shown in Fig. 20, Fig. 21, Fig. 22 and Fig. 23. The global view of the fatigue fracture surface for untreated and heat treated specimens is very similar. The process of fatigue consists of three stages – crack initiation stage (I), progressive crack growth across the specimen (II) and final sudden static fracture of the remaining cross section of the specimen (III) (Bokuvka et al., 2002; Palcek et al., 2003; Novy et al., 2007; Tillova & Chalupova, 2009; Moreira & Fuoco, 2006).

a) σ = 88 MPa, N_f = 11 560 cycles

b) σ = 54 MPa, N_f = 5.10^6 cycles

Fig. 20. Complete fracture surfaces, SEM

Stage I and II is so-called fatigue region. The three stages are directly related to the macrographic aspects of the fatigue fractures (Fig. 20). Within the bounds of fatigue tests was established that, high stress amplitude caused small fatigue region (Fig. 20a) and large region of final static rupture. With the decreasing of stress amplitude increases the fatigue region of stable propagating of cracks (Fig. 20b) and the initiation places are more focused to one point simultaneously.

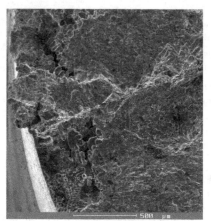

a) detail of one initiation' site on the surface

b) detail of more initiation' sites on the surface

Fig. 21. Fatigue crack nucleation - overview of a fracture surface

Important to the stress concentration and to fatigue crack nucleation is the presence of casting defects as microporosities, oxide inclusions and shrinkage porosities, since the size of these defects can by much larger than the size of the microstructure particles. It was confirmed, that if are in structure marked cast defects, then behaved preferentially as an initiation's places of fatigue damage.

The cast defects were detected on the surface of test fatigue specimens. Details of the initiations site are shown in Fig. 21. For low stress amplitudes were observed one initiation place (Fig. 21a). For high stress amplitudes existed more initiation places (Fig. 21b). The occurrence of these cast defects (Fig. 22) causes the small solubility of hydrogen during solidification of Al-Si alloys.

The main micrographic characteristics of the fatigue fracture near the initiating site are the tear ridges (Fig. 23a-c) in the direction of the crack propagation and the fatigue striation in a direction perpendicular to the crack propagation. The striations are barely seen (Figure 23d). Fig. 23b illustrates the same fatigue surface as Fig. 23a, near the initiating site, in BSE electron microscopy. The result of BSE observation presents the contrast improvement of brittle Fe-rich intermetallic phases $Al_{15}(FeMn)_3Si_2$.

Final rupture region for untreated and heat treated specimens is documented in Fig. 24. Fracture path is from micrographic aspect thus mostly transgranular and the appearance of the fracture surface is more flat. The fracture of the α-dendritic network is always ductile but particularly depends on morphology of eutectic Si and quantity of brittle intermetallic phases (e.g. $Al_{15}(FeMn)_3Si_2$ or Al_5FeSi).

The fracture surface of the as-cast samples revealed, in general, a ductile rupture mode with brittle nature of unmodified eutectic silicon platelets (Fig. 24a).

Fracture surface of heat treated samples consists almost exclusively of small dimples, with morphology and size that traced morphologhy of eutectic silicon (solution treatment resulted spheroidisation of eutectic silicon), such as those seen in Fig. 24b and Fig. 24c.

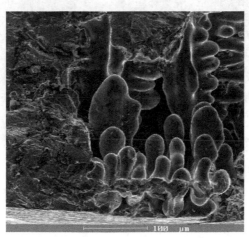

Fig. 22. Detail of cast defect

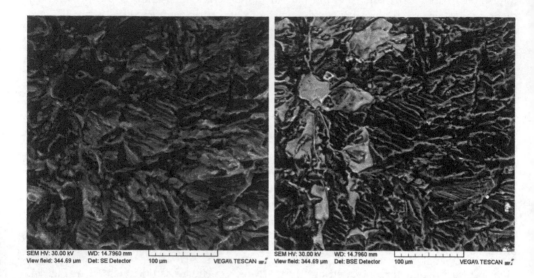

a) fatigue fracture surface near the initiating site - fine tear ridges

b) fatigue fracture surface near the initiating site - BSE

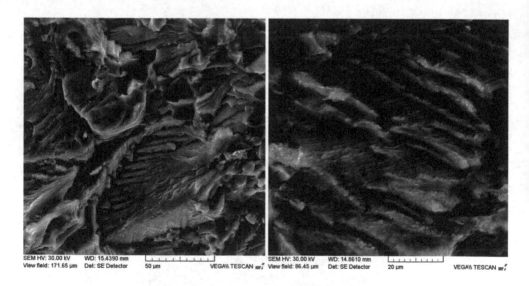

c) detail of fatigue fracture surface - tear ridges

d) detail of the typical aspect of fatigue - extremely fine striations

Fig. 23. Typical fatigue fracture surface

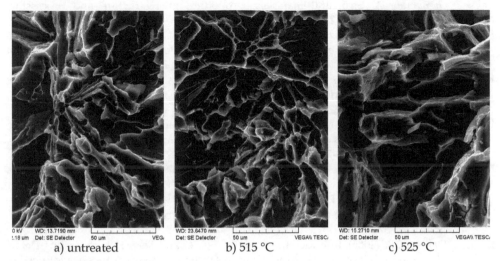

| a) untreated | b) 515 °C | c) 525 °C |

Fig. 24. Final rupture region - detail

4. Acknowledgment

The authors acknowledge the financial support of the projects VEGA No1/0249/09; VEGA No1/0841/11 and European Union - the Project "*Systematization of advanced technologies and knowledge transfer between industry and universities (ITMS 26110230004)*".

5. References

Abdulwahab, M. (2008). Studies of the mechanical properties of age-hardened Al-Si-Fe-Mn alloy. *Australian Journal of Basic and Applied Sciences*, Vol. 2, 4, pp. 839-843

ASM Handbook (1991). Volume 4, *Heat treating*. pp. 1861-1956, ASM International

Belan, J. (2008). Structural Analyses of Advanced Materials for Aerospace Industry. *Materials science/Medžiagotyra*, Vol.14, 4, pp. 315-318, ISSN 1392-1320

Bokuvka, O.; Nicoletto, G.; Kunz, L.; Palcek, P. & Chalupova, M. (2002). Low and high – frequency fatigue testing. CETRA, EDIS, Žilina, ISBN 80-8070-011-7

Caceres, C. H.; Svenson, I. L. & Taylor, J. A. (2003). Strenght-ductility Behaviour of Al-Si-Cu-Mg Casting Alloys in T6 temper. *Int. J. Cast Metals Res.*, No. 15, pp. 531-543, ISSN 1364-0461

Das, K. S. (2006). Designing Aluminum Alloys for a Recycle-Friendly World. *Materials Science Forum*, Vols. 519-521, pp. 1239-1244, ISSN 1662-9752

Das, K. S. & Gren J. A. S. (2010). Aluminum Industry and Climate Change-Assessment and Responses. *JOM*, 62, 2, pp. 27-31, ISSN 1047-4838

Dobrzański, L. A.; Maniara, R.; Krupiński, M. & Sokołowski, J. H. (2007). Microstructure and mechanical properties of AC AlSi9CuX alloys. *Journal of Achievements in Materials and Manufacturing Engineering*, Vol. 24, Issue 2, pp. 51-54, ISSN 1734-8412

Fang, X.; Shao, G.; Liu, Y. Q. & Fan, Z. (2007). Effect of intensive forced melt convection on the mechanical properties of Fe- containing Al-Si based alloy. *Materials science and engineering* A, Structural materials: properties, microstructure and processing. Vol. 445-446, pp. 65-72, ISSN 0921-5093

Krupiński, M.; Labisz, K.; Rdzawski, Z. & Pawlyta, M. (2011). Cooling rate and chemical composition influence on structure of Al-Si-Cu alloys. *Journal of Achievements in Materials and Manufacturing Engineering*, Vol. 45, Issue 1, pp. 13-22, ISSN 1734-8412

Kim, H. Y.; Park, T. Y.; Han, S. W. & Mo, L. H. (2006). Effects of Mn on the crystal structure of α-Al(Mn,Fe)Si particles in A356 alloys. *Journal of Crystal Growth*, Vol. 291, Issue 1, pp. 207-211, ISSN 0022-0248

Lasa, L. & Rodriguez-Ibabe, J. M. (2004). Evolution of the main intermetallic phases in Al-Si-Cu-Mg casting alloys during solution treatment. *Journal of Materials Science*, 39, pp. 1343-1355, ISSN 0022-2461

Li, R. (1996). Solution heat treatment of 354 and 355 cast alloys. *AFS Transaction*, No. 26, pp. 777-783

Li, R. X.; Li, R. D.; Zhao, Y. H.; He, L. Z.; Li, C. X.; Guan, H. R. & Hu, Z. Q. (2004) . Age-hardening behaviour of cast Al-Si base alloy. Materials Letters, 58, pp. 2096-2101, ISSN 0167-577X

Lu, L. & Dahle, A. K. (2005). Iron-Rich Intermetallic Phases and Their Role in Casting Defect Formation in Hypoeutectic Al-Si Alloys. *Metallurgical and Materials Transactions A*, Volume 36A, pp. 819-835, ISSN 1073-5623

Ma, Z.; Samuel, A. M.; Samuel F. H.; Doty, H. W. & Valtierra, S. (2008). A study of tensile properties in Al-Si-Cu and Al-Si-Mg alloys: Effect of β-iron intermetallics and porosity. *Materials Science and Engineering A*, Vol. 490, pp. 36-51, ISSN 0921-5093

Mahfoud, M.; Prasada Rao, A. K. & Emadi, D. (2010). The role of thermal analysis in detecting impurity levels during aluminum recycling. *J Therm Anal Calorim*, 100, pp. 847-851, ISSN 1388-6150

Maniara, R., Dobrzański, L. A., Krupiński, M. & Sokołowski, J. H. (2007). The effect of copper concentration on the microstructure of Al-Si-Cu alloys. *Archiwes of Foundry engineering*, Vol. 7, Issue 2, pp. 119-124, ISSN 1897-3310

Martinkovic, M. (2010). *Quantitative analysis of materials structures.* STU Bratislava, ISBN 978-80-2273-445-5 (in Slovak)

Michna, S.; Lukac, I.; Louda, P.; Ocenasek, V.; Schneider H.; Drapala, J,; Koreny, R. ; Miskufova, A. et. al. (2007). *Aluminium materials and technologies from A to Z.* Adin, s.r.o. Presov, ISBN 978-80-89244-18-8

Moreira, M. F. & Fuoco, R. (2006). Characteristics of fatigue fractures in Al-Si cast components. *AFS Transactions*, pp. 1-15

Mrówka-Nowotnik, G. & Sieniawski, J. (2011). Microstructure and mechanical properties of C355.0 cast aluminium alloy. *Journal of Achievements in Materials and Manufacturing Engineering*, Vol. 47, Issue 2, pp. 85-94, ISSN 1734-8412

Moustafa, M. A.; Samuel, F. H. & Doty, H. W. (2003). Effect of solution heat treatment and additives on the microstructure of Al-Si (A413.1) automotive alloys. *Journal of Materials Science*, 38, p. 4507-4522, ISSN 0022-2461

Moustafa, M. A. (2009). Effect of iron content on the formation of ß-Al$_5$FeSi and porosity in Al-Si eutectic alloys. *Journal of Materials Processing Technology*, 209, pp. 605-610, ISSN 0924-0136

Novy, F.; Cincala, M.; Kopas, P. & Bokuvka, O. (2007). Mechanisms of high-strength structural materials fatigue failure in ultra-wide life region. *Materials Science and Engineering A*, Vol. 462, No. 1-2, pp. 189-192, ISSN 0921-5093.

Palcek, P.; Chalupova, M.; Nicoletto, G. & Bokuvka, O. (2003). Prediction of machine element durability. CETRA, EDIS Žilina, ISBN 80-8070-103-2

Panuskova, M.; Tillova, E. & Chalupova, M. (2008). Relation between mechanical properties and microstructure of Al-cast alloy AlSi9Cu3. *Strength of Materials*, Vol. 40, No. 1, pp. 98-101, ISSN 1573-9325

Paray, F. & Gruzleski, J. E. (1994). Microstructure–mechanical property relationships in a 356 alloy. Part I: Microstructure. *Cast Metals*, Vol. 7, No.1, pp. 29-40

Rios, C. T. & Caram, R. (2003). Intermetallic compounds in the Al-Si-Cu system. *Acta Microscopica*, Vol.12, N°1, pp. 77-81, ISSN 0798-4545

Sencakova, L. & Vircikova, E. (2007). Life cycle assessment of primary aluminium production. *Acta Metallurgica Slovaca*, 13, 3, pp. 412-419, ISSN 1338-1156

Shabestari, S. G. (2004). The effect of iron and manganese on the formation of intermetallic compounds in aluminum-silicon alloys. *Materials Science and Engineering A*, 383, pp. 289-298, ISSN 0921-5093

Samuel, A. M.; Samuel, F. H. & Doty, H. W. (1996). Observation on the formation ß-Al₅FeSi phase in 319 type Al-Si alloys. *Journal of Materials Science*, 31, pp. 5529-5539, ISSN 0022-2461

Seifeddine, S. 2007. The influence of Fe on the microstructure and mechanical properties of cast Al-Si alloys. Literature review - Vilmer project. Jönköping University, Sweden

Seifeddine, S.; Johansson, S. & Svensson, I. (2008). The Influence of Cooling Rate and Manganese Content on the β-Al₅FeSi Phase Formation and Mechanical Properties of Al-Si-based Alloys. *Materials Science and Engineering A*, No. 490, pp. 385-390, ISSN 0921-5093

Sjölander, E. & Seifeddine, S. (2010). Optimisation of solution treatment of cast Al-Si-Cu alloys. *Materials and Design*, 31, pp. 44–49, ISSN 0261-3069

Skocovsky, P.; Tillova, E. & Belan, J. (2009). Influence of technological factors on eutectic silicon morphology in Al-Si alloys. *Archiwes of Foundry Engineering*, Vol. 9, Issue 2, pp. 169-172, ISSN 1897-3310

Skocovsky, P. & Vasko, A. (2007). *Quantitative evaluation of structure in cast iron*, EDIS Žilina, ISBN 978-80-8070-748-4 (in Slovak)

Taylor J. A. 2004. The effect of iron in Al-Si casting alloys. *35th Australian Foundry Institute National Conference*, pp. 148-157, Adelaide, South Australia

Tillova, E. & Chalupova, M. (2001). Využitie hlbokého leptania pri štúdiu morfológie eutektického kremíka. *Scientific papers of the University of Pardubice*, Series B - The Jan Perner Transport Faculty, 7, pp. 41-54, ISSN 1211-6610

Tillova, E. & Panuskova, M. (2007). Effect of Solution Treatment on Intermetallic Phase's Morphology in AlSi9Cu3 Cast Alloy. *Materials Engineering*, No. 14, pp. 73-76, ISSN 1335-0803

Tillova, E. & Panuskova, M. (2008). Effect of Solution Treatment on Intermetallic Phase's Morphology in AlSi9Cu3 Cast Alloy. *Mettalurgija/METABK*, No. 47, pp. 133-137, 1-4, ISSN 0543-5846

Tillova, E. & Chalupova, M. (2009). *Structural analysis of Al-Si cast alloys*. EDIS Žilina, ISBN 978-80-554-0088-4, Žilina, Slovakia (in Slovak)

Tillova, E.; Chalupova, M. & Hurtalova, L. (2010). Evolution of the Fe-rich phases in Recycled AlSi9Cu3 Cast Alloy During Solution Treatment. *Communications*, 4, pp. 95-101, ISSN 1335-4205

Tillova, E. & Chalupova, M. (2010). Fatigue failure of recycled AlSi9Cu3 cast alloy. *Acta Metallurgica Slovaca Conference*, Vol.1, No.2, pp. 108–114, ISSN 1335-1532

Vasko, A. & Belan, J. (2007). Comparison of methods of quantitative metallography, In: *Improvement of Quality Regarding processes and Materials*, S. Borkowski & E. Tillova, (Ed.), 53-58, PTM, ISBN 978-83-924215-3-5, Warszawa, Poland

Vaško, A. (2008). Influence of SiC additive on microstructure and mechanical properties of nodular cast iron. *Materials science/Medžiagotyra*, Vol.14, 4, pp. 311-314.,ISSN 1392-1320

Wang, Q. G.; Apelian, D. & Lados, D. A. (2001). Fatigue Behavior of A356/357 Aluminum Cast Alloys - part II. Efect of Microstructural Constituents. *J. of Light Metals*, 1, pp. 85-97, ISSN 1471-5317

Warmuzek, M. (2004a). Metallographic Techniques for Aluminum and Its Alloys. *ASM Handbook. Metallography and Microstructures*, Vol 9, pp. 711-751, ISBN 978-0-87170-706-2, ASM International

Warmuzek, M. (2004b). *Aluminum-Silicon-Casting Alloys: Atlas of Microfractographs*. ISBN 0-87170-794-2, ASM International, Materials Park

7

Catalyst Characterization with FESEM/EDX by the Example of Silver-Catalyzed Epoxidation of 1,3-Butadiene

Thomas N. Otto, Wilhelm Habicht,
Eckhard Dinjus and Michael Zimmermann
Karlsruhe Institute of Technology, IKFT,
Germany

1. Introduction

Ag catalysts are of outstanding importance in the field of heterogeneous catalysis. Optimum distribution and morphology of the Ag particles must be ensured by controlled, tailored catalyst synthesis. Hence, there is a growing demand for the characterization of Ag-dispersed fine particle systems requiring high-resolution surface observation of particles down to a few tens of nanometers and elemental analysis by *field emission scanning electron microscopy and energy-dispersive X-ray spectrometry* (FESEM/EDX). It is beneficial to characterize the particle morphology by comparison of different imaging methods like secondary electron (SE)-, backscattered electron (BSE)- and transmitted electron (TE) detection. In scanning electron microscopy surface topography becomes visible due to the dependency of the SE yield on the angle of electron incidence. Together with the large depth of field informative images of irregularly shaped particle structures are obtained. The increased BSE yield of high atomic numbers (Z) such as Ag catalysts and promoters (e.g. Cs) compared to a low-density matrix and the high penetration depth of 20-30 keV electrons also allows imaging and analysis of inclusions that would be obscured at low beam energies. Both SE and BSE detectors, in particular at low beam voltages, can additionally reveal interesting surface features of fine Ag particles. A well-known example for a Ag catalyzed reaction is the α-Al_2O_3 supported Ag-catalyzed epoxidation of 1,3-butadiene to 3,4-epoxybutene. The electrophilic addition of oxygen across the carbon-carbon double bond of 1,3-butadiene, resulting in a three-member ring structure that can undergo further chemical transformations to oxygenated products, such as ketones, alcohols, and ethers. Supported silver catalysts have been shown to epoxidize olefins with nonallylic hydrogen when an alkali promotor is doped on the surface. Thus, the direct kinetically controlled oxidation to the corresponding epoxide is preferred. The guiding hypothesis for this partially oxidation is that surface oxametallacycles are key intermediates for epoxidation on promoted Ag catalysts. Therefore, the preparative application of Ag and promoters (Cs, Ba) on the catalyst support material is of great importance. Another important aspect is sintering of Ag particles which may reduce the catalytically active surface and decreases the overall reaction performance. For this research, catalysts are produced by sequential impregnation of two mineralogically differing support materials (SC13, SLA2) with an aqueous active component solution ($AgNO_3$, $CsNO_3$). The two selected Ag catalyst systems are examined using mainly FESEM/EDX. The determination of metal amount were carried out with EDX area analysis and, if necessary, supplemented by EDX-spot analysis. In some cases, characterization of Ag-distributions by EDX mappings were made. Furthermore, for the Monte Carlo (MC)- simulations x-ray line scans of two different SEM preparation techniques (bulk specimen, thin-film supported specimen) were performed to underpin the relationships impressively. Temperature-programmed O_2 desorption (O_2-TPD) as well as N_2 sorption (BET) measurements are important analysis methods in catalysis chemistry, too. Both methods are used to support the FESEM/EDX investigation to provide complementary contributions with regard to the Ag distribution and the properties of the carrier surfaces.

2. Characterization and measurement methods

2.1 Ag catalyst systems

Huge varieties of materials are used in the preparation of heterogeneous catalysts, especially industrial catalysts. Catalysts can be divided into three groups of constituents, namely active catalytic agents, promoters, and supports [1]. Catalysts are manufactured by various methods, such as wet impregnation, leaching, drying and calcination. The major components of the catalyst system are the catalyst support (bulk material, e.g. Al_2O_3, TiO_2, SiO_2), which might influence the catalytic activity of the active components (metal-support interactions, MSI) [2] and the active metal e.g. Ag, Pd, Pt is the active agent. Increasing the surface area of the active agent is one function of the support. Maintaining a high dispersion of the active components is the other function. α-Al_2O_3 with its small specific surface area has proved to be a wear-resistant carrier material for Ag and to be highly suited for the selective oxidation of 1,3-butadiene [3].

2.2 Catalyst preparation

The Ag catalysts for this research are produced by sequential incipient wetness impregnation of two mineralogically different carrier materials with an aqueous solution of $AgNO_3$ and $CsNO_3$ as active components. The following catalyst supports are applied:

- SC13 (Almatis) mineralogical: 80 % alpha-Al_2O_3, 20 % γ-Al_2O_3, Chemical composition: 99 % Al_2O_3, 0.05 % CaO, 0.03 % Fe_2O_3, 0.02% SiO_2, 0.4 % Na_2O.
- SLA92 (Almatis) mineralogical: Main phase Ca hexa aluminate (CA6) hibonite, secondary phase alpha-Al_2O_3, chemical composition: 91% Al_2O_3, 8.5 % CaO, 0.04 % Fe_2O_3, 0.07 % SiO_2, 0.4 % Na_2O.

Dry SC13 and SLA92 are pre-sieved and subjected to wet sieving (Retsch laboratory sieves manufactured according to DIN 3310, mesh width 0.045 mm – 0.063 mm). The sieve fractions are dried at 120°C in a circulating air oven for 5 h. The desired active component solutions $AgNO_3$(aq) and $CsNO_3$(aq) are applied sequentially to the carrier materials and subjected to an ultrasound bath for 0.5 min (Sonorex RK 100H) [4]. The precursors are dried in a circulating air oven at 40°C for 5 h and then oxidized with 100 % O_2 (4.8) at 250 °C for 10 min (4000 ml_{STP} h^{-1}). A reduction with 100 % H_2 (6.0) follows at 200 °C for 10 min (1000 ml_{STP} h^{-1}). The concentrations of the active metal (oxidation number $\pm$ 0) components after conditioning are given by:

D1, SC13, 45/63 µm, 5 % Ag   MZ06, SLA 92, 45/63 µm, 5 % Ag, 1500 ppm Cs
D2, SC13, 45/63 µm, 5 % Ag, 1500 ppm Cs   MZ09, SLA 92, 45/63 µm, 10 % Ag, 1500 ppm Cs
D3, SC13, 45/63 µm, 10 % Ag
D4, SC13, 45/63 µm, 10 % Ag, 1500 ppm Cs
D5, SC13, 45/63 µm, 20 % Ag
D6   SC13, 45/63 µm, 20 % Ag, 1500 ppm Cs

2.3 Electron interactions with the specimen

Scanning electron microscopy in combination with energy-dispersive X-ray spectrometry (SEM/EDX) is a well-established and versatile method for the characterization of heterogeneous catalysts, especially Ag catalysts. They are predominantly composed of small metal particles dispersed onto a supporting material, generally a chemically inert oxide. Information about particle size distribution, deposition on the substrate surface, and elemental compositions can be obtained easily. In FESEM, a high brightness Schottky-type field emission (FE) cathode with its small beam diameter (spot size) enables imaging of features with high resolution and high contrast down to the nanoscale even on bulk substrates. Coincidently, the element composition of the specimen is available by excitation of inner shell electrons to collect characteristic X-rays with sufficient intensity for analytical information. Due to the high depth of field ($D \approx d^2 / \lambda$, $\lambda_{20\,keV} = 0.0086$ nm), where D is the depth of field, d is the apparent resolution, and λ is the electron wavelength, impressive images of differently shaped catalyst particles can be generated [5]. In the SEM, the primary electron beam creates different types of electron interactions, while penetrating the specimen. For this research, however, we utilize the following 4 important electron interactions with the specimen, namely the generation of

- secondary electrons (SE)
- backscattered electrons (BSE)
- transmitted electrons (TE)
- characteristic X-rays

Secondary and backscattered electrons are essentially for the topographical imaging of the specimen surface and therefore are described in more detail in chapter 2.4. Transmitted electrons (TE) interact less with the specimen. The specimen appears more or less "transparent" for electrons, depending on the *thickness and density* of the specimen. In energy-dispersive X-ray Spectroscopy (EDX), the X-rays are produced by inelastic scattering of primary beam electrons with bound inner shell electrons during their penetration into the matter. Subsequent deexcitation by transition of outer shell electrons to the inner shell vacancy results in emission of an element specific X-ray quantum [6]. These characteristic X-rays are essential for the determination of the elemental composition of a specimen. The electron penetration depth depends mainly on the primary beam energy and the target composition. The X-ray production range also depends on the critical excitation energy of the specific X-ray line (e.g. Kα or Lα) and is always smaller than the electron range. The electron range is the travelling distance from the primary beam electron incident at the target surface to the point where the electrons lost their energy by multiple interaction processes within the material. Figure 1 illustrates the different kinds of electron interactions with specimen and the according electron detectors.

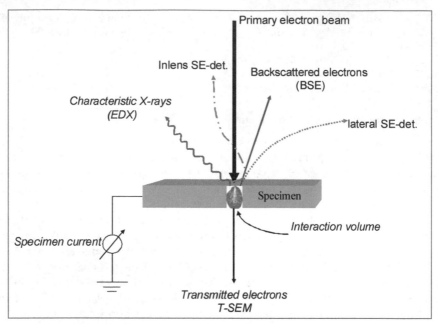

Fig. 1. Different types of electron interactions with specimen and related detection modes.

2.4 Imaging modes

Secondary electrons (SE), backscattered electrons (BSE) and transmitted electrons (TE) are responsible for three different imaging modes in SEM. The fourth type of electron interaction is the generation of X-rays and, hence, not a typical imaging mode.

The 3 imaging modes are explained as follows:

1. Backscattered electrons provide contrast based on atomic number (Z-contrast) and density. The BSE-detector will be used as a quick response to visualize materials heterogeneously composed and distributed. It will be applied often in combination with the EDX- unit to capture images for subsequent microanalysis. Depending on the primary beam energy, backscattered electrons are created inside the specimen by elastic scattering. They possess approximately the energy of the primary electrons. As a rule of thumb, the exit depth of the BSE is half the primary electron range. As a consequence, the imaging resolution achieved by backscattered electrons is worse than those achievable by secondary electrons.

2. Low energetic (< 50 eV) secondary electrons are used for true surface imaging, because they are created in the vicinity of the primary beam impact on the target surface. They are responsible for high-resolution imaging (in-lens SE detection).

3. Transmission-type images (also called T-SEM mode) are obtained by a special mounting device for TEM grids with a diode-type detector beneath. The detector is mounted like a specimen stub instead and also adjusted to the electron beam. The specimen must be sufficiently thin (< 200 nm) to permit penetration of beam electrons at 20-30 kV, which is the typical operation voltage in T-SEM mode. Imaging of mass-

thickness contrast is enabled. The application of the TE- detector can provide additional information on the sub-surface structure of many particles which cannot be resolved clearly in the corresponding SE and BSE images. The thin-film-supported specimen permits imaging with enhanced contrast due to an improved signal-to-noise ratio and X-ray analysis with reduced scattering background [7], [8].

2.5 Specimen preparation for SEM / T-SEM

The specimens are fixed with conductive tabs onto Al-stubs as bulk specimen for conventional SEM imaging. For T-SEM imaging, the samples are suspended in ethanol, and a droplet is poured onto a carbon-filmed TEM-grid (400 meshes Cu, ca. 8 nm C) for transmitted electron measurements.

2.6 SEM equipment

The electron microscopic investigations are carried out with a DSM 982 Gemini, Zeiss corp., Germany, equipped with a 4-quadrant solid-state BSE detector, a high brightness in-lens SE detector and a lateral SE detector. The DSM 982 GEMINI is applied with a thermal (Schottky-type) field-emission electron source (SFE). Element-specific quantification is performed by the dedicated EDX unit, equipped with a 30 mm^2 Si(Li) detector INCA Pentafet™, FWHM 129 eV @ MnKα (Oxford corp., England).

2.7 O_2-TPD

O_2-TPD analysis is performed using a BELCAT-B (BEL INC. Japan) system. It is coupled with a GAM 400 quadrupole mass spectrometer (In Process Instruments, Germany) as a mass-selective detector. The samples are subjected to a preliminary in-situ treatment in an O_2 / He test gas mixture. For this purpose, the sample is heated to 250°C in steps of 5 K/min and then kept at 250°C for 1 h. After this, it is cooled down to room temperature (RT). At RT, the non-adsorbed O_2 is removed by rinsing with He. In the subsequent O_2-TPD experiment the pre-treated samples are heated linearly in He. A thermal conductivity detector (WLD) is applied. The quadrupole mass spectrometer determines the composition of the desorbed products.

2.8 N_2-BET

The BET surface area (S. Brunauer, P. H. Emmett, and E. Teller) is determined with N_2 (77 K) (BELSORP-mini II, BEL INC. Japan). All samples are subjected to a preliminary treatment in a vacuum at 200°C for 5 h. To determine the sorption isotherms, the amount of molecules adsorbed on the samples are measured as a function of the relative pressure p/p_0. The specific surfaces are calculated from the adsorption parts of the isotherms at 77 K in the relative pressure range from 0.01 to 0.35 p/p_0 with N_2.

3. Discussion

3.1 Electron microscopy affecting parameters and MC simulation

FESEM and the dedicated EDX unit are important aids in the determination of the relationship among catalyst particle (cluster) sizes, dispersion onto support, support morphology and the influence of promoters. For a better understanding it makes sense to calculate the practical electron range by the Monte Carlo simulation program MOCASIM™ [9]. Additionally, the depth of the X-ray generating region (X-ray range) is estimated for the materials under consideration. It is based on an analytical expression useful for most elements and is calculated by the equation 1 of Andersen-Hasler:

$$R = \frac{0.064}{\rho} \cdot \left(E_0^{1.68} - E_c^{1.68} \right) \tag{1}$$

Whereby R (μm) is the X-ray range, E_0 (keV) is the primary electron (beam) energy, E_c (keV) is the critical excitation energy for the characteristic (analytical) X-ray line and ρ is the density of the elements. The dimension of the primary X-ray generation volume is important for the information depth obtained. It depends on the beam energy and the X-ray line chosen for the measurement [10]. The dependency of the practical electron range (penetration depth) and the X-ray range on the material density and beam voltage (beam energy) of the catalyst materials Ag, Al, and Al_2O_3 is depicted in Tab. 1. The higher the density of the element or specimen, the less is the penetration depth of the electrons (practical electron range) and the X-ray range (information depth). As a consequence, the material should be ideally homogeneous over the electron range to minimize errors caused by specimen heterogeneity. The critical excitation energy is determined by the specific absorption edge for the electron shell of an element from which the analytical X-ray line will be emitted. Therefore, E_c is always slightly higher than the corresponding X-ray line.

Specimen	ρ / g cm⁻³	X-ray line / keV	E_c / keV	HV / kV	pract. Electron range / nm	X-ray range / nm
Al	2.7	1.487 (Kα)	1.56	5	361	304
				10	1146	1085
				15	2252	2192
				20	3639	3585
Al_2O_3	3.5	-	-	5	278	235
				10	884	837
				15	1738	1691
				20	2807	2766
Ag	10.5	2.984 (Lα1,2)	3.35	5	93	45
				10	295	245
				15	579	530
				20	936	888

Table 1. Parameters affecting the electron range and X-ray range. E_c is the critical excitation energy for the corresponding analytical X-ray line. The practical (pract.) electron range is calculated by MOCASIM™.

The different penetration depths for heterogeneously composed materials will be overcome by using the T-SEM technique as it will be discussed later. As depicted in Figure 2 and Figure 3 the MC simulations elucidate the different excitation volumes for 5 keV and 20 keV beam energies. They interact with an Alox (Al_2O_3) support covered with an Ag cluster that is assumed to be 100 nm in thickness. It demonstrates the differences in the electron ranges as a function of beam energies and material density whereas the relative atomic number Ag (107.8) and the relative molecular weight Al_2O_3 (101.9) are close together. Remarkable is the beam spreading effect caused by multiple scattering in the bulk specimen, which leads to a pear-like shape of the electron trajectories (interaction volume). This will be drastically reduced by using thin supporting foils and thinning of the specimen or using sufficiently small particles, depending on the beam voltage and density of the specimen (see Figure 10).

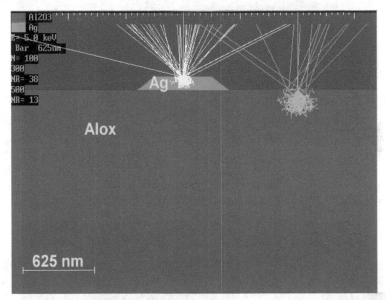

Fig. 2. MC simulation of electron trajectories (interaction volumes) created by primary beam energy of 5 keV. Note that N is equal to the number of used trajectories (100) for MC simulation. NR corresponds to the number of BS-electrons in case of Ag (38) and Alox (Al_2O_3, 13).

Obviously, specimens that are heterogeneously composed of materials strongly different in density and atomic number require a careful X-ray analysis. Nevertheless, due to the irregular shape of the analyzed particles, the systematic error of EDX measurements is increased compared to that of a homogeneous specimen with flat and smooth surfaces. Commonly, the commercially available "standard less analysis software" are adjusted to 'ideally' smooth samples. Usually, the EDX analysis results are expressed in weight-percent (wt.-%), in which the collected X-ray counts (intensity = f(element conc.)) will be converted into concentrations by means of an evaluation program taking into account fundamental X-ray parameters and the detector efficiency. The procedure is known as "standard less analysis" and is already established by the suppliers. To reduce errors resulting through different X-ray exit and scattering angles due to the different particle sizes, irregular dispersion, or agglomerated catalyst deposits, a huge population of particles should be chosen for elemental analysis of bulk specimen. To obtain e.g. the amount of metal covering the support, data collection at low magnification as demonstrated in Fig.4. should preferably be performed.

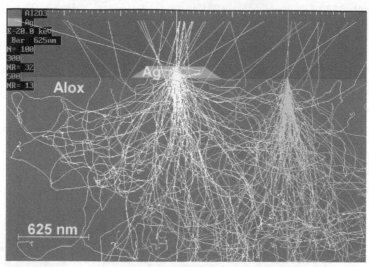

Fig. 3. MC simulation of electron trajectories created by a primary beam energy of 20 keV. Note the progress in the interaction volumes compared to Figure 2. N is equal to the number of used trajectories (100) for MC simulation. NR corresponds to the number of BS-electrons in case of Ag (32) and Alox (Al_2O_3, 13).

Fig. 4. Magnified (100 x) BSE image of a large population of Al_2O_3 particles differently covered with Ag, prepared on a conventional specimen stub.

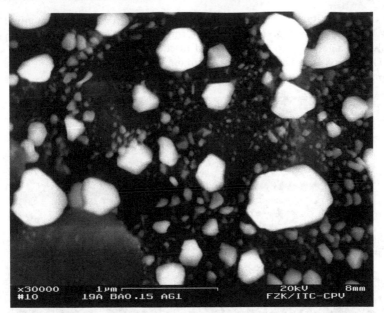

Fig. 5. High-magnification (30,000) image of a single Al₂O₃ particle covered with different sized Ag clusters, prepared on a conventional specimen stub.

Therefore a great number of particles should be included in the analyzed region to average the influence of different shaped and sized particles. As an example, a low-magnified, large number of supporting alumina (Alox = Al₂O₃) particles partially covered with Ag is shown in Figure 4 and produced under conditions mentioned in chapter 2.2. The magnification of 100 x is recommended for an EDX analysis to determine the amount of Ag deposited. The alumina particle diameter varies between 30 µm and 80 µm. For comparison, a highly magnified single alumina particle (sample D4) with differently sized and shaped Ag deposits is presented in Figure 5. It represents a high-magnification (30,000 x) image of a single substrate particle, covered with differently sized Ag clusters. The size range of the deposited Ag is between 20 nm and 850 nm.

3.2 EDX line scans

Figure 6 shows an EDX line scan which is simulated by MC calculation using a bulk specimen geometry with an Ag deposit of 100 nm in thickness onto an alumina (Alox) support. It reveals the course of Ag-Lα, Al-Kα and O-Kα X-rays across the specimen. The X-ray signals of Al and O dropped drastically down over the whole Ag cluster, contrarily the Ag signal becomes dominant. The lack in signal strength (intensity) is caused by absorption of the weak Al and O X- rays in the dense Ag particle. The maximum of the Ag signal, observed at the ends of the Ag deposit (edges) in the simulation is conspicuous. This is not observed in practice, only in the simulation.

Figure 7 is an example of an EDX line scan across the Ag agglomerate (approximately 620 nm) on a selected surface image of sample D6, which is generated by backscattered electrons. The line scan is shown in detail in Figure 8. One can see the pronounced maximum of the Ag signal and the minimum of the Al signal, whereas the minimum of the O signal is less distinct as expected. It is a hint for the presence of subsurface oxygen or Ag bulk-dissolved oxygen, which will discussed in chapter 3.6 in detail by means of O_2-TPD and BET. The comparison between the experimental EDX line scan of sample D6 (Figure 8) and the simulated line scan (see Figure 6) is feasible.

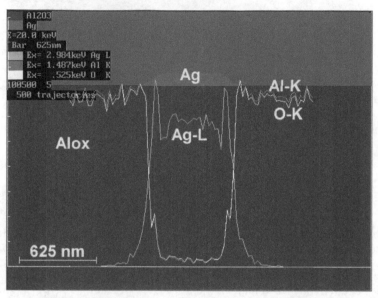

Fig. 6. MC simulation of an X-ray line scan (20 kV beam voltage) across a single particle deposited on an alumina (Alox) substrate.

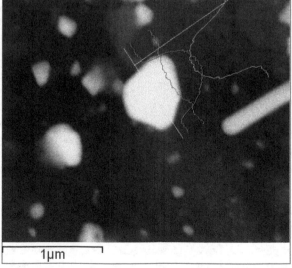

Fig. 7. BSE image of sample D6 and EDX line scan across a selected Ag deposit.

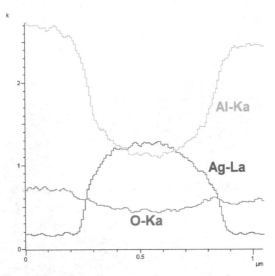

Fig. 8. Line scan in detail of Figure 7.

To demonstrate the imaging power of the T-SEM mode an example of a very tiny single alumina particle (D4) partially covered with Ag deposits in comparison to a extended particle (Figure 5) is shown in Figure 9. The dense Ag clusters appear as dark spots on the nearly "electron-transparent" alumina matrix. The graphical representation of an MC calculation based on an Al_2O_3 (Alox) substrate of 92 nm in thickness and covered with a 20 nm thick Ag deposit on a TEM-grid support is depicted in Figure 10. It illustrates the transmitted electron trajectories with a strongly reduced excitation volume in contrary to Figure 3.

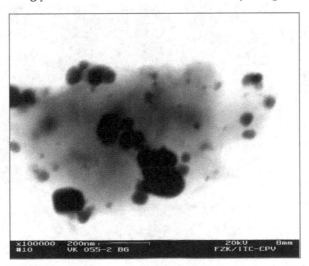

Fig. 9. Transmitted (TE) electron image (100,000 x) of an irregularly shaped single alumina particle (light grey) partly covered with Ag clusters (dark), prepared on a TEM-grid support.

The MC simulation of the electron beam trajectories in Figure 10 and the X-ray line scan in Figure 11 visualizes the interaction of primary electrons with matter of different densities and thickness on a thin-film support. The broadening of the beam (skirt) depends on the specimen thickness and density.

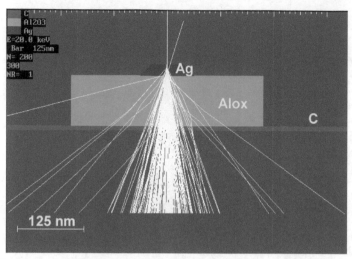

Fig. 10. MC simulation of the electron trajectories of an Ag deposit on an Alox (Al$_2$O$_3$) substrate, supported by an 8 nm carbon film (C) (TEM-grid). Note the broadening of the beam by passing the beam electrons through the matter. Backscattering events are drastically reduced (NR =1).

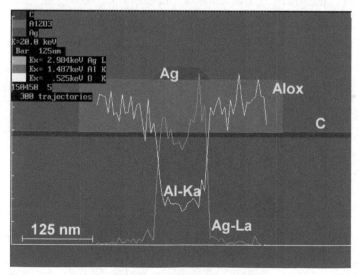

Fig. 11. X-ray line scan simulation of the thin film-supported (C) specimen. The course of the X-ray scan across the specimen is similar to that in Figure 6, but the intensity is decreased (indicated by the statistical fluctuations) due to the reduced beam interaction volume.

Thinner samples have better performance in resolution and contrast. In order to get highly magnified images even of tiny (sub nanometer) Ag particles, a TEM grid has to be prepared.

3.3 Ag distribution by EDX

In Figure 12 one can see the correlation of measured Ag concentration and the percentage of Ag area coverage visualized by image analysis (Figure 13). The image analysis is based on the intensity of the BSE signal.

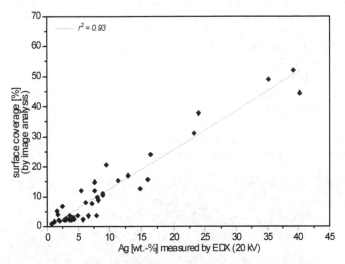

Fig. 12. Ag area coverage derived from the BSE signal and the correlation to the Ag concentration measured by EDX.

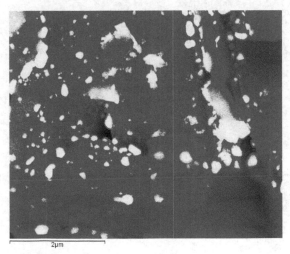

Fig. 13. Color rendering of a BSE image. The yellow areas represent Ag deposits. The red colored areas show the matrix (Al_2O_3).

The yellow spots represent the Ag deposits, approximately 13.5% from the total area. It corresponds to approximately 11 wt.-% Ag measured by EDX. This procedure will not replace the numerical calculation by the evaluation program, but visualizes in a first approximation the dependence of the Ag concentration on the BSE contrast. The Ag distribution on the carrier material is discussed with regard to the penetration depth of primary beam electrons and lower limits of detection for X-ray analysis. It is crucial to catalytic activity that Ag particles cover the carrier material almost homogeneously. Sintering of Ag particles to Ag agglomerates has to be prevented [11]. Using sample D6 (20 % Ag / 1500 ppm Cs) as an example, the FESEM images (Figure 14) and EDX analysis results (Tab. 2) shall be discussed. The image reveals uniform grain sizes of the carrier material and variable coverage (bright areas) by Ag particles or Ag clusters. In the quadrants Q1, Q2, Q3, and Q4 of the SEM image of sample D6 the coverage of the carrier grains by Ag particles (Ag-clusters) varies (bright areas). The apparently uniform catalyst grains are assumed to result from the wet sieving of the carrier material (SLA92 and SC13). The EDX spectrum in Figure 15 reveals all significant elements Ag, Al, O, and Cs. Obviously, some of the grains are densely covered with Ag, others not. Figure 16 shows the shape of huge Ag agglomerates of the D6 sample in a higher magnification (30,000 x). There is no significant difference in the measured Ag surface content between the 4 quadrants in Figure 14, which indicates an "apparently homogeneous" distribution of Ag.

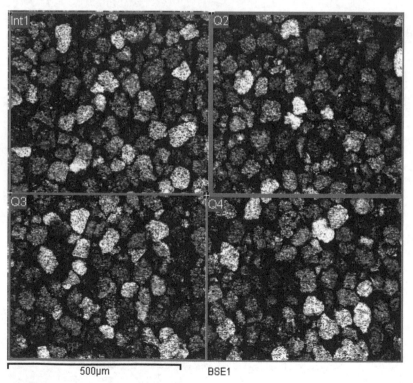

Fig. 14. SEM image of sample D6 (100 x, 20 kV) showing the 4 quadrants and the whole (integral) surface area analyzed by EDX (Tab.2).

EDX results sample D6	C	O	Al	Ag	Cs
Int1	3.0	51.8	29.8	15.1	0.12
Q1	2.6	51.3	30.2	15.9	0.00
Q2	5.4	51.2	30.3	13.1	0.00
Q3	1.9	52.2	30.6	15.2	0.15
Q4	3.1	51.4	29.1	16.4	0.00
average	3,3	51.6	30.0	15.1	0.01
standard deviation	1.3	0.40	0.60	1.30	0.07
max.	5.4	52.1	30.6	16.4	0.15
min.	1.9	51.2	29.1	13.1	0.00

Table 2. Results (wt %) of the EDX analysis of sample D6 shown in the FESEM image Figure 14 (100 x, 20 kV).

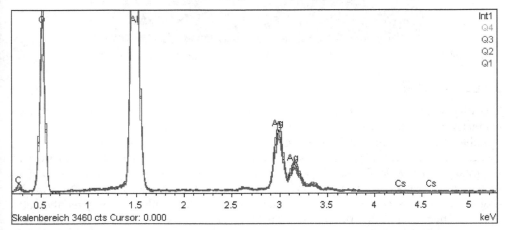

Fig. 15. EDX analysis of sample D6 (Figure 14) revealing Ag, Al, O, and Cs as significant elements.

According to Tab. 2, the average Ag concentration is 15.1% (average of Int1, Q1 to Q4) and the Ag concentrations in the quadrants do not vary significantly (Q1 = 15.9 % Ag, Q2 =13.1 % Ag, Q3 = 15.2 % Ag, and Q4 = 16.4 % Ag). But the Ag concentrations doesn't correlate well with the absolute amounts of Ag (20 wt %) used for catalyst preparation. In general, the heterogeneous catalysts are influenced considerably by the penetration depth of the electron beam because of layer thickness, Ag cluster size, surface coverage, preparation performance and, under certain conditions, Ag bulk-dissolved oxygen (see chapter 3.6). Hence, the Ag values determined by EDX could differ from the absolute Ag concentrations of the prepared samples. The accuracy is strongly influenced by morphological effects and heterogeneity of the Ag coverage of the analyzed samples. Availability of reliable standards in this respect would be beneficial. In either case, for an absolute Ag determination a quantitative digestion should be carried out (4 wt% HF) and analyzed by ICP-AES (Inductive coupled plasma - atom emission spectroscopy) under appropriate conditions.

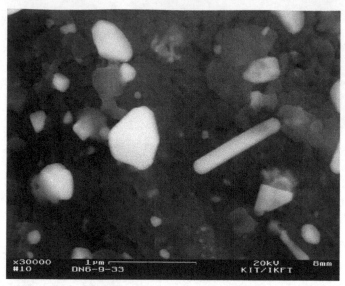

Fig. 16. Shape of huge Ag agglomerates of the D6 sample at 20 kV and 30,000 x.

3.4 EDX mapping

The dispersion of the Ag and Cs metal deposited on the catalyst support can also be characterized by the EDX mapping technique. Topographical and elemental imaging is possible at the same time. EDX mapping is a valuable tool to indicate the quality of the Ag dispersion. The electron beam is scanned pixel by pixel across a selected area of interest. In the following example, the element mappings for a selected specimen area (Figure 17, BSE image, bright spots indicating metal) revealed an inhomogeneous distribution of Ag (Figure 18) and Cs (Figure 19) of the catalyst support. Generally, the brighter the color appears, the higher is the concentration of the specific element. In both cases it correlates with the results in Tab. 3. Cs as a promoter facilitates Ag distribution on the α-Al_2O_3 surface based on the high coverage and lack of crystallites with large contact angles [12]. The presence of Cs in the catalyst improves the distribution of Ag over the support and the Ag/Al interfacial area. Area P1 (see Figure 17 and Tab. 3), for instance, shows a remarkably high silver content of 66%, which is about twice as high as the average. The deviation from the average amount of Cs is not noticeable, because Cs is only present in small quantities (ppm). Nevertheless, the distribution of Ag and Cs can be shown very impressively by the time-consuming mapping. Spectra of the analyzed regions are shown in Figure 20. In summary it can be said that the inhomogeneity of metal distribution on the catalyst carrier material can be shown very clearly also for small amounts of metal. This example demonstrates a less quality of the preparation. One can see that the texture of the substrate SLA2 is very rough in comparison to SC31, which means that a homogeneous distribution of Ag and Cs is not so easy to achieve and requires an improvement in preparation. In case of an excellent preparation, this is a optimal Ag distribution for both catalyst systems, the question is which of the two catalyst systems is more favourable for the Ag-catalyzed epoxidation of 1,3-butadiene (1,3-BD) to 3,4-epoxy-1-butene (EpB) in the reactor.

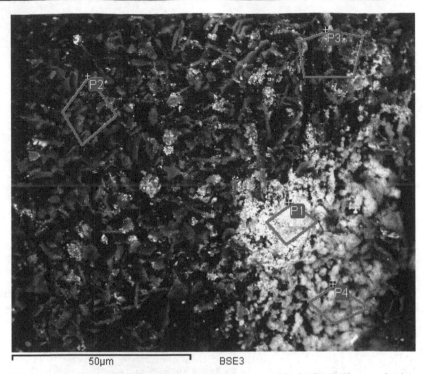

Fig. 17. BSE image of a selected area of an Ag catalyst particle (MZ06). The marked (red) areas P1, P2, P3 and P4 represent analyzed regions.

Fig. 18. Map of Ag-Lα showing the inhomogeneous distribution of Ag. The Ag concentration correlates with the intensity of the green area (measuring time about 5 h).

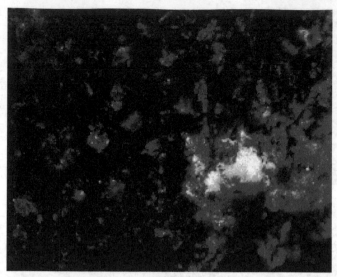

Fig. 19. Map of Cs-Lα showing the inhomogeneous distribution of Cs. The Cs concentration is correlated with the intensity of the blue area(measuring time about 5 h).

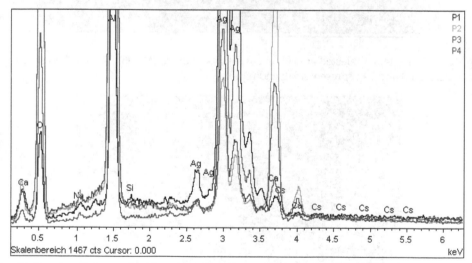

Fig. 20. Spectra of the analyzed regions (see Figure 17).

The rate of a catalyzed reaction should be proportional to the surface area of the active agent. Therefore it is desirable to have the active phase in form of the smallest possible particle. But the most undesired contribution to the reduction of the active surface (deactivation) is sintering (welding together of particles by applying heat below the melting point). The function of the support is to increase the active surface and to reduce the rate of sintering of the metal particles. On the other hand, the interaction between the lattice oxygen of the carriers and the metal particles also influences the behaviour of the active metal agent.

The support can modify the electronic character of the metal particle regarding to its adsorption and reactivity properties. Furthermore, the bond between the metal particle and the support can influence the shape of the metal particle (clusters). Both effects are so-called metal-support interactions (MSI). This effect decreases for supported Ag catalysts in the SiO_2 > Al_2O_3 > C sequence [2].

Spectrum	O	Na	Al	Si	Ca	Ag	Cs	Sum
P1	21.8	0.45	10.08	0.15	0.67	**66.33**	0.45	100
P2	54.1	0.32	31.16	0.00	6.62	**7.60**	0.18	100
P3	52.5	0.18	34.09	0.08	3.59	**9.53**	0.00	100
P4	38.6	0.28	19.59	0.04	2.16	**38.32**	0.92	100
average	41.8	0.31	23.73	0.07	3.26	**30.45**	0.39	100
max.	54.1	0.45	34.09	0.15	6.62	**66.33**	0.92	
min.	21.8	0.18	10.08	0.00	0.67	**7.60**	0.00	

Table 3. All results in wt.-%, derived from the BSE image in Figure 17.

Another tool to elucidate the distribution of elements is the so called CAMEO™ imaging. Figure 21 represents a CAMEO™ image from the examined specimen which is based on BSE detection. CAMEO™ is a tool to convert X-ray energies into visible wavelengths. In comparison to the X-ray mapping technique, the CAMEO™ procedure is much faster, but it may lead in some cases to a false color rendering (color overlap) caused by adjacent X-ray energies. As depicted in Figure 21 one can see that the brownish areas obscured the small Cs-spots. Therefore, Cs and Ag are not distinguishable because they are located at the same area of surface.

40µm

Fig. 21. CAMEO™ rendering of the area shown in Figure 17. The green colored areas represent Ag, whereas the brownish areas represent Al, respectively.

3.5 Remarks

All analytical results have been obtained at 20 kV acceleration voltages, which turned out to be the best choice for excitation conditions. Because the in-lens detector is switched off above 20kV, therefore high resolution SE imaging is disabled. The EDX results are determined by the manufacturer's spectrum evaluation software (Oxford corp.). Originally, all quantitative results are calculated with two significant fractional decimal digits provided with 1σ errors, which includes the errors resulting from spectra processing (background subtraction, filtered least squares fitting, peak overlap). Notably, this may not reflect uncertainties caused by surface topography and other systematic influences. According to experience, the actual uncertainties are considerably higher, especially for light elements. We suggest for C and O a relative uncertainty of 5 – 20%, for all other elements 1-5% [13]. Another important issue is the estimation of the *limit of detection* (LOD), which can be calculated from a synthetic spectrum by the equation 2.

$$LOD_{3\sigma} = 3 \cdot \sqrt{B} \cdot \frac{C}{P} \qquad (2)$$

Where B is the number of background counts, $3 \cdot B^{1/2}$ represents 3σ error of background measurement, C is the concentration of the element, P corresponds to the number of counts in the X-ray line after background subtraction. In practice, calculation will be performed by a special program tool named '*spectrum synthesis*', which is provided with the INCA-Energy evaluation software [14].

LOD 3σ / w t.- %		I / nA	t_m / s	U / kV
Ag	Cs			
0.21	-	0.23	60	20
0.18	0.27	0.23	100	20
0.09	0.15	0.71	100	20
0.06	0.09	0.71	300	20

Table 4. Calculated 3σ LOD's regarding to equation 2 and [14]. The matrix composition, except for alumina, is assumed to be 1.8 wt.-% Ca, 0.3 wt.-% Na, 0.2 wt.-% P.

For the 3σ LOD estimation, the composition of the sample matrix and the acquisition parameters like measuring time t_m, beam voltage U, current I, detector parameters, X-ray take-off angle are required. Tab. 4 shows 3σ LOD's which are calculated for 3 different measuring times and aperture adjustments (beam current). For Ag, this is of minor importance, since the amounts of Ag are sufficiently high in comparison to the promoters.The pore size of the alumina support is in a similar order of magnitude as the X-ray generating range (see Fig. 1b), which leads to an uncertainty in the measurements in these regions. As a consequence, the EDX analysis is considered to be semi-quantitative on such Al_2O_3 supports. A loss of X-rays (Al-Kα and O-Kα) emitted from mesoporous media compared to that of dense monocrystalline alumina is described in literature [15]. It may be caused by charge effects due to the specific surface of particles and distribution of pores. In the course of our various measurements, no significant charging was observed. Compared to common *inductively coupled plasma atomic emission spectroscopy* (ICP-AES) the values of metal deposition determined by EDX are frequently higher. One explanation is a sometimes observable incomplete chemical digestion prior to ICP measurements. In the case of Al_2O_3 we recommend a digestion with HF (4%). As already mentioned, EDX with an electron penetration depth in the order of several nanometers up to microns is a technique for surface analysis, while the ICP technique is applicable to the quantitative determination of the bulk composition with the detection limits in the µg/l (ppb) range. Due to the incipient wetness impregnation technique applied here for surface preparation, the values of metal deposition analyzed by EDX are expected to be higher than those by ICP-AES [16]. The differences in the amounts of Ag are plausible. The reason therefore is that the D-samples and MZ-samples are completely chemically digested, which is a physical homogenization. In contrast, the sample preparation for FESEM/EDX is non-destructive, which means that the sample is not physically homogenized. Tab. 5 gives a selected overview about catalysts characterization methods applied on sample D1.

Promoter CsNO$_3$ / wt.-%	ICP Ag / wt.-%	EDX Ag / wt.-%	BET m^{2}g^{-1}
0.15	4.46	5.76	6.03

Table 5. A listing of commonly applied laboratory methods (ICP, EDX, BET) for characterization of a catalyst with nominal 5 wt.-% Ag, grain fraction 45-63 µm (sample D1).

3.6 O$_2$-TPD and BET measurement

To get a deeper insight it makes sense to combine SEM with O$_2$-TPD experiments and BET measurements, which additionally were carried out. The TPD experiment measures the temperature-dependent desorption rate of a molecule from the catalyst surface [17]. Typically, O$_2$ / Ag interactions are studied on the Ag monocrystal surfaces (110) and (111) [18]. This means that silver may storage O$_2$, with the amount adsorbed being dependent on temperature and O$_2$ partial pressure. O$_2$ dissociatively adsorbs on the Ag catalyst and may assume the following characteristic forms:

- Surface oxygen
- Subsurface oxygen
- Ag bulk-dissolved oxygen

Ag bulk-dissolved oxygen may act as a storage of converted surface oxygen and, hence, be supplied later on. Subsurface O$_2$ increases the coordination number of Ag surface atoms, which results in a smaller binding strength of surface oxygen and favourably influences the epoxidation reaction of 1,3-BD. The surface area has been corrected by the subsurface oxygen value. Comparison of the measured amount of O$_2$ desorbed with the amount theoretically required for an O$_2$ monolayer shows that the measured amount of desorbed O$_2$ is higher by a factor of 2. This may be explained by the presence of subsurface O$_2$ [19].

O_2-TPD studies reveal significant differences between MZ samples with an SLA92 carrier and D samples with an SC13 carrier for the same Ag contents. In general it can be said that the desorbed O_2 amounts of the MZ samples are below those of the D samples by a factor of 5. Moreover, the temperatures of the desorption maximums of the MZ samples are lower. This difference (binding strength of oxygen) results in a strongly variable reactivity of the adsorbed oxygen species. Figure 22 shows O_2 desorption and temperature of the samples D1, D3, and D5 with different Ag concentration on the support that is *not doped with Cs*. With increasing Ag concentration, the desorptions maxima are shifted towards lower temperatures (red line). This indicates that epoxidation already may start at lower temperatures. Figure 23 shows that the sample D6 doped with Cs reaches a higher temperature at the desorption maximum than the non-doped sample D5.

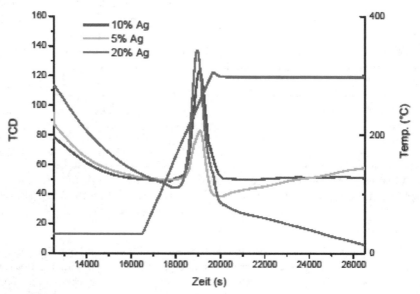

Fig. 22. O_2 detection (WLD) by O_2-TPD of the samples D1 (5 % Ag), D3 (10 % Ag), and D5 (20 % Ag) versus temperature (red line).

This indicates that the adsorbate is stabilized by the presence and the grade of distribution of Cs and Ag on the surface, which correlates with TPD results, FESEM, EDX mapping and the reaction performance. The determination of the specific surface areas of the catalysts D3, D4, D5, D6, MZ06 and MZ09 gives an important hint regarding to metal distribution of MSI effects. Tab. 6 lists the specific surface areas for the mentioned catalysts determined with N_2 as adsorptive. All isotherms are of the IUPAC type II ("s-shaped") and, hence, can be evaluated according to the BET theory [20], [21]. The Cs-doped samples (D4, D6) have a slightly higher surface than the non-doped samples (D3, D5) with the same Ag content, which correlates to a higher desorption rate for O_2-TPD. In case of the samples MZ06 and MZ09 one can see that the difference in BET surface is not remarkable, even in the presence of different Ag amounts. The total error for BET measurement is about 0.1 m²/g.

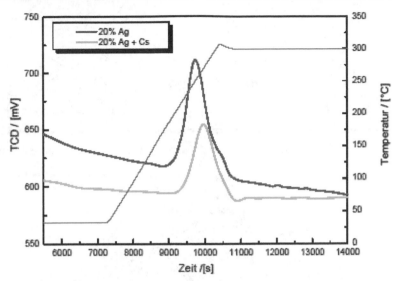

Fig. 23. O2 detection (WLD) by O2-TPD of samples D5 (20 % Ag) and D6 (20 % Ag + Cs) versus temperature (red line).

	D3	D4	D5	D6 Ag 20% +	MZ06	MZ09
	Ag 10%	Ag 10% + Cs	Ag 20%	Cs	Ag 5% + Cs	Ag 10% + Cs
$a_{s,BET}$ [m^2 g^{-1}]	6,0424	6,5521	7,6083	8,0461	0,80722	0,7009

Table 6. Specific surface areas from sample D3 – D6 and MZ06 and MZ09.

The BET reveals that there is a significant difference in the catalyst surface, which is expressed by the topological images of FESEM of the D samples (SC31) and MZ (SLA92) samples. Note: Very small amounts (0,1g) of sample are used for the investigations (FESEM, TPD). In the case of BET-Measurement the sample weight is in the range of 0.1 g up to 10 g, depending on the specific surface. The carrier material SC13 (high BET Surface) was not found to be suited for the epoxidation of 1,3-butadiene (Figure 24) on Ag particles. Already at 180°C a total oxidation of 1,3-butadiene to CO_2 and H_2O does occur, also when Cs-doped catalysts are used. In contrast to this, epoxidation of the Ag catalyst with SLA29 (less BET surface) carrier material results in an EpB selectivity of 74 % (200° C, SV = 2590 h^{-1}) at a 1,3-BD conversion rate of 15 % [22]. The constitution of the oxametallacycle intermediate is depicted in Figure 25. The guiding hypothesis is that surface oxametallacycles are key intermediate for epoxidation on Ag / Cs catalysts. The intermediate EpB(ads), finally leading to molecular EpB, is probably strongly adsorbed on the catalyst surface indicated by theoretically calculations which also support its identity as an oxametallacycle. The oxametallacycle intermediate is more thermodynamically stable than EpB by approx. 24 kcal/mol. Moreover, the transition state for EpB formation from the oxametallacycle intermediate is actually lower in energy than the reactants, butadiene and oxygen [23].

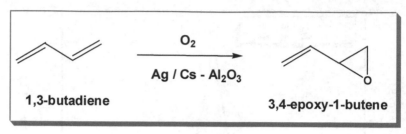

Fig. 24. Epoxidation of 1,3-butadiene over Ag / Cs catalyst on Al_2O_3 to 3,4-epoxy-1-butene.

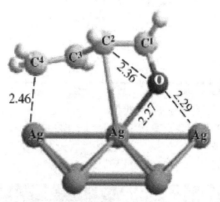

Fig. 25. The "intermediate EpB" finally leading to molecular EpB. The oxametallacycle intermediate is more thermodynamically stable than EpB. Moreover, the transition state for EpB formation from the oxametallacycle intermediate is actually lower in energy than the reactants, butadiene and oxygen

4. Conclusions and outlook

Production of Ag catalysts based on a corundum-containing (SC13) and a calcium hexa aluminate-containing (SLA92) carrier material is crucial to the selective epoxidation of 1,3-butadiene. Optimum distribution and morphology of the Ag particles must be ensured by controlled, tailored catalyst synthesis. An increase in the activity by enhanced Ag dispersion on a corundum-containing carrier material with a larger surface area leads to completely unselective catalysts. FESEM/EDX results provide major information with regard to the Ag distribution and the properties of the carrier surfaces. The carrier material SC13 was not found to be suited for the epoxidation of 1,3-butadiene on Ag particles. At 180°C already does a total oxidation of 1,3-butadiene to CO_2 and H_2O occur also when Cs-doped catalysts are used. In contrast to this, epoxidation of the Ag catalyst with SLA29 carrier material results in an EpB selectivity of 74 % (200 °C, SV = 2590 h^{-1}) at a 1,3-BD conversion rate of 15 %. All the analysis methods complement each other to form an overall impression, which is reflected in the product selectivity, catalyst activity and educts conversion during the reaction. The distribution and composition of metal particles on the surfaces can be seen and detected with EDX and FESEM. Also topological and morphological effects can be shown. BET measurements allow drawing a conclusion for successful metal loading. Furthermore, the efficiency of Ag-loading and promoters with TPD measurements can be determined. The analytical result reflects the final behaviour of the epoxidation regarding to the product selectivity and conversion rate of the educts.

5. Acknowledgement

Doreen Neumann-Walter, (preparation), KIT, Germany
Bernhard Powietzka, (preparation, reaction control), KIT, Germany
Sara Essig, (preparation), KIT, Germany
Dr. Angela Puls, (BET measurement), Rubotherm GmbH Bochum, Germany
Dr. Volker Hagen (O_2-TPD measurement), Rubokat GmbH Bochum, Germany

6. References

[1] R. J Wijngaarden, A. Kronberg, K. R. Westerterp, Industrial Catalysis, WILEY-VCH Verlag GmbH Weinheim, 1998

[2] B. Cornils, W. A. Hermann, R. Schlögl, Chi-Huey Wong, Catalysis from A to Z, WILEY-VCH Verlag GmbH, 2000

[3] J. R. Monnier, *Prepr. Pap. - Am. Chem. Soc.,Div. Fuel Chem.* 2007, 52 (2), 163

[4] M. Pohl, S. Hogekamp, N. Q. Hoffmann, H. P. Schuchmann, Chem. Ing. Tech. 2004, 76 (4), 392

[5] J. M. Thomas, R. M. Lambert (Ed.), Characterisation of Catalysts, Chichester, Wiley 1980

[6] P. L. Gai, E. D. Boyes, Electron Microscopy in Heterogeneous Catalysis, Inst. of Physics Publishing, Bristol a. Philadelphia, 2003

[7] W. Habicht, N. Boukis, G. Franz, O. Walter, E. Dinjus, Microsc. Microanal. 2006, 12, 322-326

[8] C. E. Lyman et al.; Scanning Electron Microscopy, X-ray Microanalysis, Analytical Electron Microscopy: a Laboratory Workbook; Plenum Press N. Y. (1990)

[9] L. Reimer, Monte Carlo Simulation of Electron Diffusion, updated Version 3.1, Program and Handbook, Münster, Germany 1998

[10] J. I. Goldstein et al., Scanning Electron Microscopy and X-ray Microanalysis, 3rd ed., Kluwer Acdemic/Plenum Publishers, New York, 2003

[11] Handbook of Heterogeneous Catalysis, Vol.1, 2nd Ed. (Eds: G. Ertl, H. Knözinger, F. Schüth, J. Weitkamp) Wiley-VCH, Weinheim 2008, 561

[12] D. M. Minahan, G. B. Hoflund, W. S. Epling, D. W. Schoenfeld, J. of Catalysis 168, 1997, 393-399

[13] W. Habicht, N. Boukis, E. Hauer, E. Dinjus, X-ray Spectrometry 2011, 40, 69-73

[14] INCA Energy, tools for INCA users, provided by Oxford corp., or more profound: P. Duncumb, I. R. Barkshire, P. J. Statham, Microsc. Microanal. 2001, 7, 341-355

[15] L. Sorbier, E. Rosenberg, C. Merlet, X. Llovet, Mikrochim. Acta, 2000, 132, 189-199

[16] C. Xu, J. Zhu, Nanotechnology 2004, 15, 1671-1681

[17] M.I. Szynkoska, E. Lesniewska, T. Paryjczak, Pol. J. Chem. 2003, 77, 657

[18] F. M. Leibsle et al., Phys. Rev. Lett. 1994, 72, 569, 569-2572.

[19] G.W. Busser, O. Hinrichsen, M. Muhler, Cata. Lett. 2002, 79 (1 - 4), 49

[20] D. D. Do, Adsorption Analysis: Equilibria and Kinetics, Imperial College Press, London
     1998
[21] P. Christopher and S. Linic, ChemCatChem 2010, (1), 2, 78
[22] T. N. Otto, P. Pfeifer, S. Pitter, B. Powietzka, *Chem. Ing. Tech.* 2009, 81 *(3)*, 349
[23] Mark A. Barteau, Topics in Catalysis Vol. 22, Nos. 1/2, January 2003

Part 2

Nanostructured Materials for Electronic Industry

8

A Study of the Porosity of Activated Carbons Using the Scanning Electron Microscope

Osei-Wusu Achaw

Department of Chemical Engineering, Kumasi Polytechnic, Kumasi,
Ghana

1. Introduction

The earliest mention of the significance of porosity in the performance of activated carbons is generally attributed to the French chemist Antoine-Alexandre-Brutus Bussy who in a 1822 publication suggested that porosity was important to the adsorptive properties of activated carbons. Since then a lot of research has gone into elucidating the nature of porosity of activated carbons, its development and measurement. In particular, a great deal of research has been spent on understanding factors that affect the development of porosity and how to model the porosity in terms of these factors. Similarly, much effort has gone into identifying accurate methods and procedures for characterizing activated carbons in general and particularly its pore structure. The continued interest in these research is because of the continued use and importance of activated carbons in industry and an unrelenting pursuit to improve on its performance. Characterization of porosity is often done indirectly by measurement of secondary data from which the requisite pore parameters are estimated. But direct methods also exist for characterizing the pore structure of activated carbons. Methods such as optical microscopy and scanning electron microscopy (SEM), in view of their ability to directly view the micro-structure of activated carbons have demonstrated enormous potential for use in the study and characterization of activated carbons [Manocha et al., 2010; Lazslo et al., 2009; Achaw & Afrane, 2008]. However, this latter approach has only been applied in a very limited capacity in the past. Rather, industry and researchers alike continue to rely on the indirect methods to determine and quantify porosity in activated carbons. The indirect methods calculate activated carbon characteristics from measurement of other parameters that are generally thought to relate to the properties of interest. Adsorption measurements and related mathematical models wherein information regarding the pore structure of an activated carbon is determined are the most commonly used amongst the indirect methods. Porosity measurements using this approach extracts such pore characteristics as pore volume, surface area, pore size distribution and average pore diameter based on mathematical models of the adsorption process, information on the adsorbate and an adsorption isotherm. Besides adsorption measurements, several other indirect methods also exist to estimate the pore characteristics of activated carbons. Among these are immersion calorimetry, small angle scattering of X-rays (SAXS), small angle scattering of neutrons (SANS), and mercury porosimetry[Rigby & Edler, 2002; Stoeckli et al, 2002; Daley et. al., 1996].

The weakness of the indirect methods, is that they are based on models that do not always match with observed behavior of activated carbons. Others like mercury porosimetry are based on very simplified descriptions of the pore structure of activated carbons that are greatly deviated from pores observed directly using direct methods. Not surprisingly, pore characteristics estimated based on two different such models or methods rarely agree [Rodriguez-Reinoso and Linares-Solano, 1989]. The weaknesses notwithstanding, the indirect methods have thus far served a useful purpose of providing a framework for assessing and comparing activated carbons. In particular, they have provided a useful vehicle for predicting and evaluating the performance of these materials for industrial and other applications. The drawbacks of these methods, however, have meant that more consistent and reliable methods continue to be searched to measure the characteristics of activated carbons. The direct methods represent a viable option in that regard. Direct methods allow the direct viewing of the topography of the activated carbon surfaces which makes possible improved description of activated carbon properties such as pore shapes and pore orientation. When coupled with other methods or instrumentations, such as computerized image analysis, it is possible to estimate the pore characteristics of activated carbons more accurately. Again, these methods make possible a visual follow up of the stages of activated carbon manufacture which in turn makes possible the tracking of the changes that a precursor material goes through in forming an activated carbon. It thus offers enormous possibilities of shedding light on the pore development processes than hitherto known [Achaw & Afrane, 2008]. Already, in areas such as materials engineering, biology and medical sciences, the SEM has been extensively used to study and characterize the microstructure of substances [Chira et al., 2009; Vaishali et al., 2008; Chung, et al., 2008; Kamran, 1997]. The purpose of this chapter is to discuss the potential use of the SEM in understanding porosity development in activated carbon and pore structure characterization using micrographs of coconut shells at different stages during the manufacture of coconut shell-based activated carbons.

2. Tracking porosity development using the scanning electron microscope

To better control porosity in activated carbons, it is essential that its development during its preparation be well understood. It is now generally known that porosity in activated carbons is derived from three main sources, namely, the inherent cellular structure of the precursor material, the conditions extant during the preparation of activated carbons and the composition of the precursor material [Heschel & Klose, 1995; Raveendran et al., 1995; Evans & Marsh, 1979;]. How these factors combine to produce an activated carbon of a given specification has been and continues to be a subject of intense research. This continued search is borne out of the need to find newer applications for activated carbons and an unending desire to improve on the performance of activated carbons in such operations like filtration, gas and metal adsorption and separation, gas storage, and finally in water purification. In all these applications the performance of an activated carbon depends as much on the total pore volume as it does on the pore size distribution, the prevalence of a certain pore size regime, and the surface chemistry of the carbon. For instance, during operations involving molecular sieve activated carbons, that pore size characteristics is required that permits the separation of two or any number of molecules of differing molecular sizes. Achieving this kind of performance demands a special design of the pore structure of activated carbons. This in turn demands not only an understanding of how porosity in activated carbons is developed but an additional insight into how to control its development. The different pore sizes play unique roles during activated carbon application. Indeed, the classification of the pores in activated carbons into micropores, mesopores and macropores is based more on the varied behavior of admolecules in these pore regimes than on the actual sizes of the pores. Thus, more than the total pore volume or total surface area of the activated carbons, the fraction of the total pore volume or surface area due to the various sizes of pores is of utmost importance. Again, understanding this development is essential for the design of models to describe the performance of activated carbons and the prediction of activated carbon behavior. According to IUPAC nomenclature [Sing, et al., 1985], micropores are those pores with width less than 2 nm. The micropores play the key role of providing the bulk of the surface involved in adsorption, which is the basis of many applications of activated carbons. The mesopores are wider than the micropores and have pore widths in the range 2 nm to 50 nm. The mesopores also play a role in adsorption albeit on a reduced scale compared to the micropores. The role of the mesopores becomes more important during the adsorption of large molecules that cannot be accommodated in the micropores. Finally, there are the macropores which have much larger pore sizes and which play the important role of being the conduits through which access to the interior of the activated carbon and hence to the mesopores and micropores are achieved. They are generally considered as being part of the external surface of the activated carbon. The macropores have size greater than 50 nm.

Pores in activated carbons are areas of zero electron density in the carbon matrix. These constitute volume elements distributed throughout the particle and posses varied sizes and shapes. The individual volume elements are connected with each other through open channels which are themselves also volume elements. The volume elements are now known to originate from several sources. First, there are those whose source can be traced directly to the primary pore structure of the precursor material. Another group of pores are created as a result of the imperfections that arise from the arrangements of the lamellar constituent molecules (LCM) which are the building blocks of activated carbons. The LCM are layers of sheets which are made up of interconnected aromatic rings. They are formed when the precursor materials are subjected to heat treatment at the appropriate temperature and conditions [Bryne & Marsh, 1995; Evans & Marsh, 1995]. The imperfect arrangements of the LCM creates space in-between parallel layers of the molecules. Volume elements are also created when parts of the LCM are reacted away during contact with agents used in the activation process. The volume elements arising as a result of LCM arrangements and reactions constitute microporosity, and to a lesser extent mesoporosity in the activated carbon. The microporosity confers on activated carbons the unique ability to adsorb large quantities of a diverse range of molecules which makes activated carbons so useful in separation processes and other applications.

The LCM and the accompanying microporosity are formed when the original cellular structure of the precursor material undergoes molecular transformation and reconstitution. During heat treatment of the precursor material a number of physical and chemical processes occur that culminate in the final activated carbon pore structure. Among these, first, moisture and other volatile constituents of the precursor material escape leaving voids that may be later transformed or retain themselves in the final activated carbon product. Secondly, the macromolecules of the precursor material breakdown, lose mostly oxygen and hydrogen and reconstitute into aromatic rings which become the building blocks of the LCM. The new constituent molecules form to enclose the vacancies left by the escaping elements and molecules. These vacant lots also constitute porosity in activated carbons. The transformations are initiated during the pyrolysis of the precursor material and are continued and enhanced during the subsequent activation stage. The loss of volatile matter from the precursor materials occurs at all temperatures but aromatization of the material occurs at temperatures in excess of 700ºC. Another important process occurs during the activation process to create new pores or enhance existing pores formed during the pyrolysis step. During activation, activation agents react with the carbon skeleton to create new pores or enlarge existing ones. It is also at the activation stage that other phenomena that facilitate pore creation manifest. For instance, inherent mineral matter such as alkali metals in the precursor material catalyze the pore formation process leading to such phenomena as pitting, channeling and pore enlargement [Bryne & Marsh, 1995]. Pores are also developed as a result of thermal stress on the cellular structure of the precursor material. This stress leads to the development of cracks, crevices, slits, fissures, and all manner of openings in the matrix of the ultimate carbon material. The events leading to the formation of pores occur mainly at the micro and sub micro levels and most of the products of the process such as LCM and associated carbon rings are hardly, directly, observable even with the most powerful of electron microscopes available today. As such these processes have most remained in the realm of theoretical discourse. However, there are other manifestations of these transformations that with the appropriate tools are observable. Using the SEM it has been possible to view images of some of the phenomena that engender porosity development in activated carbons. The ability of the SEM to distinguish objects as small as 1 nm makes it ideal for tracking the transformations happening in the precursor material during activated carbon formation.. This SEM has however not been fully exploited yet for the study of porosity development in activated carbons safe for the pioneering work of Achaw & Afrane, 2008. Figures 1-3 below, show SEM micrographs of sections of coconut shell at different stages during the preparation of coconut shell-based activated carbons. The images reveal details about these materials that shed useful light on aspects of porosity development in activated carbons.

Micrographs in Figures 1 – 3 reveal details of activated porosity development that, previously, has only being a matter of theoretical discourse. Samples for the SEM micrographs used in this study were prepared by cutting sections of well dried coconut shell (raw coconut shells, carbonized shells or coconut shell-based activated carbon) and mounting a on specimen stub with the help of a conductive silver adhesive. The specimen surfaces were thereafter sputter coated with a thin film of silver, placed in the sample holder and viewed with a Ziess DSM 962 electron microscope. A look at Figures 1 and 2 suggest that the original cellular structure of the coconut shells as seen from the transverse section are largely maintained albeit in a modified form following carbonization. In Figure 1 the largely isolated cylindrical units, see positions labeled A and B, has walls made up of layers of thin sheets. In Figure 2, these units seen in Figure 1 have joined together at the walls into a singular solid matrix interspersed with pores. The sheets of the walls are no longer visible in Figure 2. The joining of the walls and the fusing together of the sheets of the walls suggests a profound transformation at the molecular level in the shell during the pyrolysis.

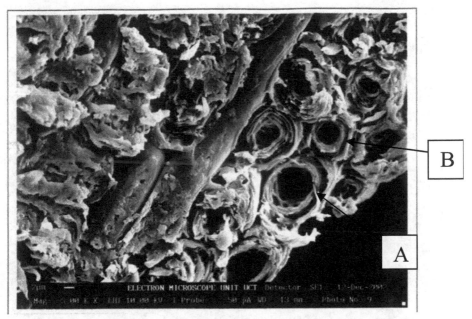

Fig. 1. Micrograph of surface of transverse section of raw coconut shell.
Source: *Achaw & Afrane, 2008*

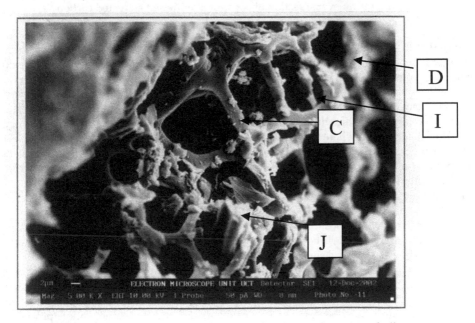

Fig. 2. Micrograph of surface of transverse section of carbonized coconut shell
Source: *Achaw & Afrane, 2008*

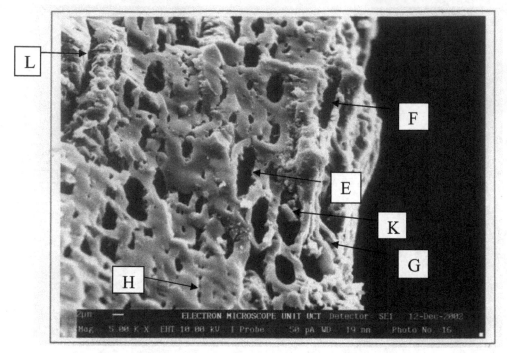

Fig. 3. Micrograph of transverse section of coconut shell-based activated carbon
Source: *Achaw & Afrane, 2008*

Notice that the cylindrical units in Figure 1 has deformed into all manner of shapes after the pyrolysis, see Figure 2, positions marked C and D. The transformation of the pores continues to the activation stage, see positions marked E and F in Figure 3, where what were originally cylindrical shape has now become partially flattened cylinder. The micrograph of the activated carbon further demonstrates transformation of the matrix of the shell as a result of continued heating during the activation. The narrowing of the pore widths suggest a kind of deformation where the matrix softens and walls of the pores give way and close in on each other. Notice further that as a result of this deformation, some of the walls have completely fallen in on each other resulting in a complete zipped up of the pores, see positions marked G and H. The transformations suggest that the carbon matrix passes through a plastic phase as a result of the thermal treatment. Another phenomena observable from these micrographs is the preponderance of foreign materials in the pores of the carbonized product, see positions marked I and J on Figure 2, and the almost lack of these materials in the activated carbon of Figure 3. This means that the activation process serves the additional purpose of cleaning foreign materials from the carbon besides the creation of pores. It is nonetheless noteworthy that even at the activated carbon stage some pores still remain blocked by foreign material, see position marked K in Figure 3. This last observation is an indication that the activation process was not complete. It is anticipated that these foreign materials would be completely cleared at the end of the activation process. Yet another important feature of these micrographs is the position marked L on Figure 3 which is a crack in the carbon matrix probably developed as a result of thermal stress on the carbon matrix due to temperature changes in the activation process. Such cracks contribute to the overall surface of the activated carbon and as such are important. Yet discourse on porosity development of activated carbons often ignore these cracks.

3. Characterization of activated carbons

Characterization of activated carbons is driven by the need to have qualitative and quantitative information which serve as the basis for comparison and selection of activated carbons for specific applications. Such data are also useful for modeling the behavior and performance of activated carbons. Furthermore, characterization provides feedback for use in the design and preparation of activated carbons. The characteristics often measured are density, abrasion resistance, surface area, average pore size, pore size distribution, pore shape, pore volume, and the surface chemistry of the carbon. Whilst there are well established standard methods for measuring the density and abrasion resistance, scientists and industry are still grappling with what accurate methods to use for measuring the others. Most current methods estimate these parameters indirectly from measurements of secondary data on the activated carbons. As a result there are still concerns with the accuracy of values determined for these parameters. The most popular method for characterizing activated carbons is through the measurement of adsorption data and application of mathematical models that relate the adsorption data to such characteristics as pore volume of the adsorbent and the properties of the adsorptive. Other indirect methods, namely, mercury porosimetry, immersion calorimetry, small angle scattering of X-rays (SAXS), neutrons (SANS), high-resolution transmission electron microscopy are also sometimes used to determine the characteristics of activated carbons. Then there are direct methods that hold enormous potential for characterizing activated carbons but which use are rarely mentioned in activated carbon literature. These latter methods are mainly the microscopic methods which enable the observation of micro- and sub-micro features of activated carbons and hence the direct measurement of these features. These microscopic methods are optical microscopy and the SEM.

3.1 Adsorption methods of characterizing activated carbons

These methods almost invariably combine the adsorption isotherm of a given adsorbate-adsorbent system and a theoretical or empirical model of the adsorption process to estimate the characteristics of activated carbons [Machnikowski, et al., 2010; Noor & Nawi, 2008; Lozano-Castello et al., 2004; Stoeckli et al., 2002; Stoeckli et al., 2001 Yuna et al., 2001; Rodriguez-Reinoso, 1989]. Most commercial sorption equipment estimate activated carbon data using in-built software based on one version or the other of this approach. Of historical importance is the Langmuir model [Gregg & Sing, 1982] which was first developed in 1916 to describe adsorption behavior on solid adsorbents in general. The model relates the adsorption of molecules in a gaseous medium onto a solid surface to the gas pressure above the solid at a fixed temperature and can be expressed mathematically as

$$\frac{P}{V} = \frac{P}{V_m} + \frac{1}{bV_m}$$ (1)

V is the equilibrium adsorbed amount (mmolg^{-1}) of the adsorbate per unit mass of the adsorbent at a pressure P. V_m is the amount of gas required for monolayer coverage of the adsorbent (mmolg⁻¹), and b is a constant whose value depends on the temperature. A linear plot of equation (1) allows V_m to be evaluated from the gradient and hence the adsorbents surface area from the relation

$$S = V_m L \sigma \tag{2}$$

S is the total surface area of the adsorbent (m²g⁻¹), L is Avogadro's number and σ is the projected surface area of the adsorbate molecule. S is the sum total of pore surfaces and external (non-pore) surface of the adsorbent. Equation (1) is based on the assumption that i) there is a mono-layer adsorption, ii) there are no adsorbate-adsorbate interactions on the adsorbent surface, iii) the adsorbent has a homogeneous surface, iv) all adsorption sites on the adsorbent are equivalent and, v) the adsorbing gas adsorbs into an immobile state.

The Langmuir's model has been found to be of limited applications for activated carbons. In particular, activated carbon surfaces are rarely homogeneous. Generally, the assumptions have been found not to be consistent with observations therefore the Langmuir model is rarely used to characterize activated carbons. Consequently, other relatively more accurate models of adsorption are often used. One such model is the Braunnauer, Emmett and Teller (BET) model [Sing et. al., 1985]. The BET method has a much wider application and is most often used to interpret adsorption isotherms obtained using Nitrogen at 77K as the adsorbate. The model is an improvement on the Langmuir model in that it can account for multilayer adsorption. It relates the adsorption pressure and the volume of the adsorbed adsorbate according to the equation

$$\frac{p}{V(p^o - p)} = \frac{1}{V_{mc}} + \frac{c-1}{V_{mc}} \frac{p}{p^o} \tag{3}$$

where

$$c = \exp(\frac{(\Delta H_A - \Delta H_L)}{RT}) \tag{4}$$

In equation (3), V_{mc} is the monolayer capacity of the adsorbent, pᵒ is the saturation vapour pressure of the adsorbate gas, p is the pressure of the gas, and c is a constant which is exponentially related to the heat of first layer adsorption. ΔH_A is the heat of adsorption, ΔH_L is the heat of liquefaction of the adsorption fluid, T is the temperature, and R is the gas constant. A linear plot of equation (3) allows V_{mc} to be determined from the intercept and from which the surface area of the activated carbon can be estimated using equation (2). The BET method has found a number of applications in adsorption studies and is especially used in the determination of the surface areas of adsorbents including activated carbons. Even so, the BET equation is unable to account for adsorption in a number of instances. For activated carbons equation (3) is only linear at p/pᵒ < 0.1. This introduces error into the measurement of those pores for which adsorption is possible at pressures for which p/pᵒ > 0.1. Secondly, the calculation of the adsorbent surface area using equation (2) requires knowledge of the projected surface area of the adsorbate molecule, in this case Nitrogen. This in turn requires that the adsorbate molecules (Nitrogen) be in a close packed, monolayer coverage on the adsorbent. The use of the method implicitly assumes that the value estimated for V_{mc} is necessarily accurate and that σ is constant for the adsorbate under all conditions. Further, adsorption in micropores is characterized more by pore filling than by surface coverage. As such, the application of the BET method does not always yield the correct result for the surface area of activated carbons, especially if they are predominantly microporous.

Probably more accurate among the adsorption methods for the determination of pore characteristics of activated carbons is the Dubnin-Raduskevitch (DR) equation and its improved and more versatile version, the Dubnin-Astakov (DA) equation [Dubnin & Raduskekevich, 1947; Dubnin, 1989; Carrasco-Merin et al., 1996; Gil, 1998]. The DR equation is premised on the assumption that adsorption in micropores occurs by pore filling rather than by physical adsorption on the surface of the micropores. The equation relates the volume of pores, W, filled by an adsorbate at a given temperature T and relative pressure p/p^o and other parameters of the adsorption system as

$$W = W_o \exp[-(\frac{A}{\beta E_o})^2]$$
(5)

where W_o is the total volume of the micropores, Eo is the characteristic energy, and β is the affinity coefficient. Both Eo and β are system dependent. The deferential molar work of adsorption on the adsorbent, A, is further defined as

$$A = RT\frac{p}{p^o}$$
(6)

For slit-shaped pores, a relationship exists between E_o and the average micropore width $\overline{L}$ as

$$\overline{L}(nm) = 10.8(E_O - 11.4)$$
(7)

The DR equation has a narrow range of application as it corresponds to mostly Type I isotherms. The Dubnin-Astakov (DA) equation which is a modification of the DR equation and which is applicable to a wide range of microporous carbons is therefore preferred. The generalized form of the DA equation is

$$W = W_o \exp[-(\frac{A}{\beta E_o})^n]$$
(8)

When n=2, equation (8) becomes equal to the DR equation. Values of n between 1 and 4 are observed for most carbon adsorbents, n> 2 for molecular sieve carbons or carbons with highly homogeneous and small micropores, n < 2 for strongly activated carbons and heterogeneous micropore carbons. For monodisperse carbons, n=3 and for strongly heterogeneous carbons n= 2 [Carrasco-Merin et al., 1996].

3.2 Characterization of activated carbons with the scanning electron microscope

In the areas of porosity development and characterization of activated carbons, a number of issues still remain unresolved. The current state of knowledge has not been able to address all observed behavior and performance of activated carbons. For instance, to what extent does thermal stresses on the carbon matrix during pyrolysis affect porosity development.

How realistic is the often used slit-shaped pore model in describing activated carbons. Similarly, in characterizing activated carbons the often used methods such as adsorption measurement and mercury porosimetry all rely on secondary data and mathematical models to estimate pore characteristics. But these methods are fraught with a number of drawbacks. In these methods, only those pores can be characterized that the adsorbate molecules could have access to. Also the mathematical models of adsorption which are the basis of estimating pore characteristics are based on assumptions most of which do not match with observations. Adsorption measurements in particular have other drawbacks that affect the accuracy of parameters estimated. For instance, phenomena such as activated diffusion and gate effects introduce errors into adsorption based estimates of pore parameters. Equally, idealized pore models such as the cylindrical or slit-shape pore which are the basis of a number of methods for estimating average pore width of activated carbons are too simplistic in the face of the observed complex nature of porosity in activated carbons. Then again some of the pore parameters are not at all amenable to estimation by the indirect methods. Such parameters like pore shape, pore location and distribution, and pore orientation have all eluded estimation by the indirect methods. These parameters nonetheless have important consequences on modeling and prediction of performance characteristics of activated carbons and therefore are worth estimating or measuring.

Direct methods, particularly, microscopy offers an alternative approach to resolving most of the drawbacks of the indirect methods. Using microscopy, it is possible to observe the micro and submicro-features of activated carbons directly and therefore makes possible a proper qualitative and quantitative description of its characteristics [Ito & Aguiar, 2009; Daley et al., 1996; Tomlinson, et al., 1995; Hefter, J., 1987; Ball & McCartney, 1981]. There are two types of this method, namely, optical microscopy which has a resolution of about 1 μm and electron microscopy whose resolution is much greater and in the range of about 1.5 nm and which can achieve a magnification of about 2,000,000x. The SEM is one version of the electron microscopy which uses a beam of electrons to scan the surface of a specimen and makes possible the direct observation of its surface features at the micro and submicro levels. Due to the huge magnifications and impressive resolutions achievable with the SEM it has been used in many areas of science and industry, particularly in materials engineering, biological and medical sciences for the study and characterization of the micro-structure of substances. It use provides an avenue to resolve some of the yet unresolved issues in activated carbon porosity development and characterization. The SEM functions exactly as its optical counterparts except that it uses a focused beam of electrons instead of light to "image" a specimen and gain information about its structure and composition. The SEM can yield information about the topography (surface features of an object), morphology (shape and size of the particles making up the surface of an object), composition (the elements that the object is composed of and the relative amounts of these) and crystallographic information (how the atoms are arranged in the object). This ability makes the SEM a hugely useful instrument for the study of activated carbons. The topographic information attainable using the SEM allows that surface features such as pore characteristics, the description of which has been a major preoccupation of activated carbon chemists, to be studied and measured directly from SEM micrographs. Also SEM's ability to reveal compositional details of a specimen makes it a potent instrument for studying the surface chemistry of activated carbons. Other features of the SEM that make it a unique instrument for studying activated carbons include its ability to reveal details of a sample less than 1 nm in size. This means that, in principle, even micropores could be exposed for study by SEM micrographs. Again, due to the very narrow electron beam employed, SEM micrographs have a large depth of field that yields a pseudo three-dimensional appearance useful for understanding the surface structure of a sample. Figures 4-7 are examples of SEM micrographs that reveal details of activated carbons that only the direct methods can show.

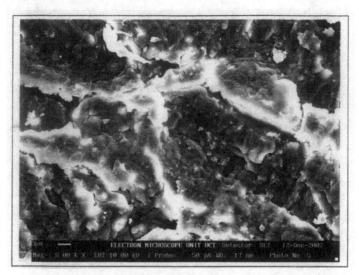

Fig. 4. Micrograph of outer surface of raw coconut shell
Source: *Achaw & Afrane, 2008*

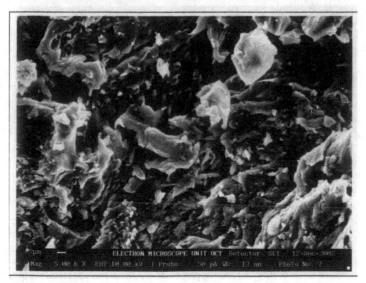

Fig. 5. Micrograph of outer surface of carbonized coconut shell.
Source: *Achaw & Afrane, 2008*

Fig. 6. Micrograph of outer surface of coconut shell-based activated carbon
Source: Achaw & Afrane, 2008

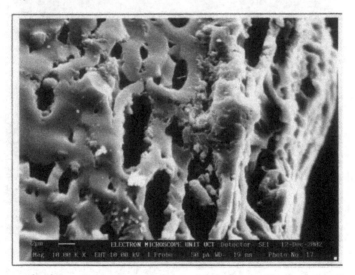

Fig. 7. A Micrograph of transverse section of coconut shell-based activated carbon
Source: This study

The micrographs in Figures 4-7 are all of coconut shells at different stages during the preparation of activated carbons. Figures 4-6 are micrographs of the surfaces of the outer sections of the shell. The surface features observed from these micrographs are significantly different from those seen from the corresponding micrographs of on Figures 1-3, indicating that the nature of pores seen of activated carbons depend on the sections used. Crucially, the nature of porosity observed in Figure 6 is totally different from any description of porosity previously described except in the work of Achaw and Afrane (2008). What is observed here is a network of cracks which are more the result of thermal stress on the carbon matrix. This is totally in contrast to what was observed of the micrograph in Figure 3 which is more likely the result of re-arrangement of molecules in the carbon matrix. The observed porosity in Figure 6 further show the difficulty of defining a generalized pore structure for activated carbons. It calls to question such often used concepts as average pore size, and pore size distribution. Further, it raises questions about models associated with pore structure such as the slit-shaped model which is the basis of a number of mathematical models of adsorption in activated carbons or the cylindrical models used by mercury porosimetry and other empirical methods for estimating pore characteristics. Figures 6 & 7 are micrographs of different sections of an activated carbon. The two images clearly demonstrate the extent of inhomogeneity of the surfaces of activated carbons. Equally noteworthy of the micrographs in Figures 4-6 is the fact that there is hardly any common trend linking the structures in these micrographs. Whilst, hardly, any pore is observable at all in Figure 4, the structure in Figure 5 is fuzzy and confusing, and hardly yielding to any definition at all. Finally, even though the structure in Figure 6 has some semblance of order, it also defies any exact definition. An important observation of these micrographs is that pore development in activated carbons is as a result of several phenomena. Particularly, it seems that thermal stress plays an important role in pore development in these materials than previously thought. The foregoing observations demonstrate the strength of SEM in studying activated carbons.

4. The scanning electron microscope

The main features of the SEM are an electron source which provides the electrons that interact with the material to be examined, an arrangement of metal apertures, magnetic lenses and scanning coils or deflectors plates that confines, focuses and turns the beam of electrons into a thin and focused monochromatic beam which is accelerated towards the sample and which irradiates the specimen in a raster fashion [Goldstein, J et al., 2003; Reimer, L, 1998]. The interaction of the electrons with the specimen initiates a number of reactions inside the sample which results in the generation of signals which are taken advantage of to gain information about the sample. The SEM imaging process involves four major steps. These include sample preparation, the specimen scanning process, image formation and image analysis. The kind of preparation required of the sample depends on whether it is electrically conducting or not. Electrically conducting samples, for instance metals, only require minimal sample preparation prior to mounting on a sample stub for scanning and imaging. Non-conductive specimens such as activated carbons, however, must first be made conducting before mounting for study. Otherwise, these tend to charge when scanned by the electron beam leading to scanning faults and other image artifacts. Non-conducting samples are therefore first sputter coated with an ultra-thin coating of an electrically-conducting material before imaging. Other reasons for coating the sample surface are to increase the signal and surface resolution, especially with samples of low atomic number. Some of the commonly used materials for coating samples include gold, graphite, platinum, chromium, tungsten, osmium, and indium. For biological materials it is possible to increase the conductivity without coating by impregnating them with osmium before imaging. It is also possible to image non-conducting specimen without coating by using the Environmental SEM(ESEM) or the field emission gun (FEG) SEM [Schatten & Pawley, 2008; Hardt, T. A. 1999;]. Samples for SEM study do not need to be as thin as it is the case in optical microscopy or transmission electron microscopy (TEM). The specimen size is dictated primarily by the size of the sample chamber and must be rigidly mounted on the specimen stub. Again, in view of the applied vacuum, it is important that the sample be completely dry. Hard, dry materials such as activated carbons can be examined with little further treatment besides sputter coating.

The signals of interest in the characterization of activated carbons using the SEM are the secondary electrons whose detection and imaging gives information on the surface topography and hence on the pore structure of the activated carbons, the backscattered electrons and the X-ray radiation which give complementary information of the chemical composition of the sample surface. Whilst secondary electrons can be detected and imaged in a conventional SEM instrument, harnessing signals from the backscattered electrons and X-ray radiation requires complementary instrumentation. The secondary electrons are produced when an incident electron in the primary electron beam excites an electron in an atom of the sample and loses most of its energy in the process. The excited electron moves to the surface of the sample where it can escape if it possesses sufficient energy. In view of their low energy, only secondary electrons that are very near the surface (<10 nm) can exit the sample and be detected. The secondary electrons are detected and accelerated onto a photomultiplier from which an amplified electrical signal output is displayed as a two-dimensional intensity distribution that can be viewed and photographed on an analogue video display or converted and displayed or stored as a digital image.

Backscattered electrons (BSE) are produced when the primary electron beam hits the sample and some of the electrons are reflected or scattered back out of the specimen. The production of backscattered electrons varies directly with the specimen's atomic number. High atomic number elements backscatter electrons more strongly than low atomic number elements, and thus appear brighter in an image. BSE are therefore used to detect contrast between areas on the sample surface with different chemical compositions. This provides opportunity for examination of the chemistry of the surface of activated carbons. Dedicated backscattered electron detectors, usually either scintillator or semiconductor types are positioned above the sample to detect the backscattered electrons.

As earlier intimated, inelastic scattering, places the atoms of the sample in an excited state. The tendency therefore is for an atom to return to its ground or unexcited state. To achieve this, the atom gives off the excess energy. This may result in the production of X-rays, cathodoluminescence and Auger electrons. The relaxation energy is the fingerprint of each element in the sample. Thus detection and analysis of the relaxation energies enables the identification of the specific elements in the surface of a sample. When an SEM is equipped with energy-dispersive X-ray spectroscopy (EDX) or wavelength dispersive X-ray spectroscopy (WDS), it is possible to get information on the elemental composition on the surface of the specimen. For an activated carbon, this method can be used to study the effect on porosity development of specific elements in the carbon and generally the chemistry of the activated carbon surface [Afrane & Achaw, 2008].

4.1 3D images in scanning electron microscope

In conventional (standard) SEM a pseudo-three dimensional (3D) view of the sample surface can be observed directly. The standard SEM image is, however, really a two-dimensional (2D) structure from which mostly qualitative data is possible regarding the microstructure of the specimen. In the study of activated carbons however, qualitative data though useful as for instance in understanding pore development phenomena, it is quantitative data on pore structure- pore sizes, pore shapes, pore surface area, and pore size distribution- that are most useful for characterization and modeling of performance and behavior. Getting quantitative data from 2D SEM images of activated carbons, however, poses a number of challenges. First, 2D SEM cannot determine three-dimensional porosity of the materials because the conventional SEM is unable to observe images of inner parts of a specimen. Then also there are the difficulties associated with getting precise descriptions or a representative pore structure in view of the otherwise complex, varied and numerous pores in the field of view of the microscope. To get quantitative data of a specimen from conventional SEM images, the practice is to convert the 2D images into 3D from which the requisite quantitative data can be measured or estimated often with the help of computerized image analysis software. A number of methods for getting 3D data from 2D images are available [Marinello et al., 2008; Spowart, 2006; Spowart et al. 2003; Alkemper and Voorhees, 2001; Lyroudia et al., 2000]. These include stereology, photogrammetry, photometric stereo, and the more useful serial sectioning method. An automated variation of the serial sectioning method called the focus ion beam-scanning electron microscopy (FIB-SEM) is increasingly being used in the areas of materials engineering and biological sciences to study the micro-features of substances. This last method appears to have enormous potential for use in the study and characterization of activated carbons.

In serial sectioning, 2D images of a sample are collected after a series of successive layers of equal width have been removed from the sample. Afterwards the stack of 2D data files (2D SEM images) is combined and processed in such a way that the microstructural features that are within the 3D data stack can be classified. This is most efficiently done using computerized image processing software. Serial sectioning is made up of two basic steps that are iteratively repeated until completion of the experiment. In the first step a nominally flat surface of the sample is prepared using any of a variety of methods such as cutting, polishing, ablating, etching, or sputtering. These processes remove a constant depth of material from the specimen between sections. The second step is to collect 2D characterization data after each section has been prepared, for instance by imaging with an SEM. Finally, computer software programs are used to construct a 3D array of the characterization data that can be subsequently rendered as an image or analyzed for morphological or topographical features of the sample. Using this method any micro and submicro features of the sample that can be distinguished by the SEM can also be characterized. Thus the capability of the methodology in characterizing topographical and morphological features of substances is limited only by the resolution of the SEM.

Automated serial sectioning techniques which facilitates the methodology exist for material removal and imaging. An example is the FIB-SEM which combines ion beam sectioning with SEM imaging to generate tomographic data that are well suited to characterize microstructural features of a sample in 3D via serial sectioning [Desbois et al., 2009; Orloff et al. 2003]. While less common, the FIB-SEM method has demonstrated the ability to complete 3D volumetric reconstruction at a resolution of 10 nm or better in all three dimensions. The method has been widely used for studies in materials engineering and life sciences and holds tremendous potential for characterizing the porosity of activated carbons.

5. Conclusion

In spite of the tremendous progress in development and use of activated carbons a number of questions still remain that the conventional indirect methods of studying activated carbon are still grappling to answer. The mode of porosity development is one such area. Another area is the characterization of the porosity of activated carbons, where the existing mathematical models and methods have still not succeeded in finding accurate ways to estimate pore parameters. Recent developments in scanning electron microscopy, especially in the conversion of 2D SEM to 3D and computerized image analysis has opened avenues for improved study of porosity development and characterization of activated carbon. This potential of the SEM has not really been adequately explored for the study of activated carbons as not much work exits in the literature in that regard. However, judging from the enormous strides researchers in the areas of materials engineering and the biological sciences have made in using this methodology to identify micro and submicro-features of substances, it is anticipated that its adaptation for use in the study of activated carbons would facilitate the study of porosity development and pore characterization. The SEM micrographs shown in this work clearly demonstrate this point. The major limitation of the SEM is the level of resolution achievable with it currently. At 1.5 nm, this poses difficulty in characterizing most micropores in activated carbons. It is nonetheless hoped that continued advances in SEM instrumentation will overcome this difficulty and facilitate the use of the SEM in the study of activated carbons.

6. References

Achaw, O-W. & Afrane, G. (2008). The evolution of pore structure of coconut shells during the preparation of cocnut shell-based activated carbons. *Microporous and Mesoporous Materials*, 112, pp. 284 – 290, ISSN 1387-1811

Afrane, G. & Achaw, O-W. (2008). Effect of the concentration of inherent mineral elements on the adsorption capacity of coconut shell-based activated carbons, *Bioresource Technology*, Vol. 99, No. 14, pp. 6678 – 6682, ISSN 0960-8524

Alkemper, J., Voorhees P. W. (2001). Quantitative serial sectioning analysis. *Journal of Microscopy*, Vol. 201, No.3, pp. 388–394, *ISSN: 1365-2818*

Ball, M. D. & McCartney, D. G. (1981). The measurement of atomic number and composition in a SEM using backscattered detectors. *J. Microsc.* 124, 57-68, ISSN: 1365-2818

Bryne, J. F. & Marsh, H. (1995). Origins and structure of porosity, in John W. Patrick (ed), *Porosity in carbons: Characterization and applications*, Edward Arnold, London, ISBN 0 340544732

Carrasco-Marin, F., Alvarez Merino, M. A., & Moreno-Castilla, C. (1996). Microporous activated carbons from a bituminous coal. *Fuel*, Vol. 75, No. 8, pp. 966-970, ISSN: 0016-2361

Chung, W., Sharifi, V. N., Swithenbank J, Osammor, O., Nolan, A. (2008).. Characterisation of Airborne Particulate Matter in a City Environment. *Modern Applied Science*, Vol.2, No. 4, p.17, ISSN 1913 – 1844

Dalye, M. A., Tandon, D., Economy, J. abd Hippi, E. J. (1996). Elucidating the porous structure of activated carbon fibers using direct and indirect methods. *Carbon*, Vol. 34, No. 10, pp 1191-1200, ISSN: 0008-6223

Desbois, G, Urai, J. L., & Kukla, P. A. (2009). Morphology of the pore space in claystones – evidence from BIB/FIB ion beam sectioning and cryo-SEM observations. *eEarth Discuss.*, 4, 1–19, 2009, www.electronic-earth-discuss.net/4/1/2009/

Dubnin, M. M. (1989). Fundamentals of the theory of adsorption in micropores of carbon adsorbents: Characteristics of their adsorption properties and microporous structures. *Pure & Applied Chemistry*, Vol. 10, pp. 1841 – 1843, ISSN 0033-4545

Evans, M. & Marsh, H. (1979). Origins of microporosity, mesoporosity, and macroporosity in carbons and graphites, in *Characterization of Porous Solids*, Unger, K. K., Rouguerol, J., Sing, K. S. W., Kral, H., (eds). London, SCI, ISBN 0 - 444 – 42953 –

Farkas, L., Major, N., Mihalko, A., Abraham, J. & Kozar, Z. (2009). Analysis of active carbon catalysis. *Material Sciences and Engineering, Miskolc*, Volume 34/2 , pp 41-51

Gil, A. (1998). Analysis of the micropore structure of various microporous materials from nitrogen adsorption at 77K. *Adsorption*, Vol. 4, Numbers 3-4, pp 197-206

Goldstein, J., Newbury D. E., Joy, D. C, Lyman, C. E., Echlin, P., Lifshin, E., Sawyer, L., and Michael, J. R. (2003). *Scanning Electron Microscopy and X-Ray Microanalysis.* Kluwer Academic /Plenum Publishers, New York, ISBN 978-0-306-47292-3

Gregg, S. J. and Sing, K. S. W. (1991). *Adsorption, Surface Area and Porosity.* Academic Press, 2nd Edition, London, 1991, ISBN 0123009561

Hardt, T. A. (1999). Environmental SEM and related applications. In Impact of electron and scanning probe microscopy on materials research. (eds). Rickerby, D. G. Giovanni Valdrè, G., & Valdrè, U, Kluwer Academic Publishers, Netherlands, p. 407, ISBN 0-7923-5939-9

Hefter, J. (1987). Morphological characterizations of materials using low voltage scanning electron microscopy. *Scanning Microsc.* 1(I), 13-21, ISSN: 1932-8745

Heschel, W. & Klose, E. (1995). On suitability of agricultural by-products for the manufacture of granular activated carbon. *Fuel*, Vol. 74, No. 12 , ISSN: 0016-2361

Ito, L. X. & Aguiar, M. L. (2009). A study of the porosity of gas filtration cakes. *Brazilian Journalof Chemical Engineering.* Vol. 26, No. 02, pp. 307 – 315. ISSN 0104-6632

Joshi, V. C., Khan, I. A., &. Sharaf, M. H. M. (2008). Use of Scanning Electron Microscopy in the Authentication of Botanicals. *Pharmacopeial Forum. Vol. 34, No.4, ISBN 0363-4655*

Kamran, M. N. (1997).Fracture analysis of concrete using scanning electron microscopy. *Scanning* Vol. 19, pp. 426 - 430, ISSN: 1932-8745

Lozano-Castello, D., Cazorla-Amoros,D., Linares-Solano, A. (2004).Usefulness of CO_2 adsorption at 273 K for the characterization of porous carbons .*Carbon 42*, pp. 1231–1236, ISSN: 0008-6223

Lyroudia, K, Pantelidou, O. Mikrogeorgis, G., Nikopoulos, N., & Pitas,L. (2000). Three-dimensional reconstruction: A new method for the evaluation of apical microleakage. *Journal of Endodontics, Vol. 26, No. 1, pp. 36- 38*, ISSN: 0099-2399

Machnikowski, J., Kierzek, K., Lis, K., Machnikowska, H., and Czepirski, L. (2010).Tailoring porosity development in monolithic adsorbents made of KOH-activated pitch coke and furfuryl alcohol binder for methane storage. *Energy Fuels* , 24, 3410–3414,

Marinello, F., Bariani, P., Savio, E., Horsewell, A., & De Chiffre, L. (2008). Critical factors in SEM 3D stereo microscopy. *Measurement Science and. Technology.* Vol. 19, No. 6, pp. 1- 12, ISSN 0957-0233

Md Noor., A. A. B. & Nawi, A. B. M. (2008).. Textural characteristics of activated carbons prepared from oil palm shells activated with ZnCl2 and pyrolysis under nitrogen and carbon dioxide. *Journal of Physical Science*, Vol. 19(2), 93–104, 2008, ISSN: 1675-3402

Mfanacho, S. M. , Hemang, P. & Manocha, L. M. (2010). Enhancement of microporosity through physical activation. PRAJÑĀ - *Journal of Pure and Applied Sciences*, Vol. 18, pp. 106 – 109, ISSN 0975 – 2595

Orloff, J., Utlaut, M., Swanson, L., (2003). *High Resolution Focused Ion Beams: FIB and Its Applications*. Kluwer Academic/Plenum, New York, ISBN 0-306-47350-X

Raveendran, K., Ganesh, A., Khilar, K. C. (1995). Influence of mineral matter on boimasspyrolysis characteristics. *Fuel* Vol. 74, No. 12, ISSN: 0016-2361

Reimer, L. (1998). *Scanning Electron Microscopy: Physics of Image Formation and Microanalysis*, Springer-Verlag, New York, ISBN 3-540-63976-4

Rigby, S. P. & Edler, K. J. (2002). The influence of mercury contact angle, surface tension, and retraction mechanism on the interpolation of mercury porosimetry data. Journal of Colloid and Interface Science 250, pp. 175-190

Rodriguez-Reinoso, F. & Linares-Solano, A. (1989). In 'Chemistry and Physics of Carbon', . P. A. Thrower (ed), Dekker, Vol. 21 , New York, p.1, ISBN 0824781139

Rodriguez-Reinoso, F., An overview of methods of the characterization of activated carbons. *Pure & Applied Chemistry*, Vol. 61, No. 11, pp 1859 – 1867, 1989, ISSN 0033-4545

Schatten, H & Pawley, J. (2008). *Biological low voltage field emission scanning electron microscopy. Springer Science + Business Media, New York*, e-ISBN 978-0-387-72972-5

Sing., K. S. W., Everett, D. H., Haul, R. A. W., Moscou, L., Pierotti, R. A., Rouquerol, J. & Siemieniewska. (1985). Reporting on physiosorption data for gas/solid systems with special reference to determination of surface and porosity. *Pure & Applied Chemistry*, Vol 57, p 603, ISSN 0033-4545

Spowart, J. E. (2006). Automated serial sectioning for 3-D analysis of microstructures. *Scripta Materialia*, Vol. 55, No.1, pp. 5–10, ISSN: 1359-6462

Spowart, J. E., Mullens H. M., Puchala, B. T. (2003) Collecting and analyzing microstructures in three dimensions: A fully automated approach. *JOM Journal of the Minerals, Metals and Materials Society*, Vol. 55, No. 10, pp. 35–37, ISSN 1047-4838

Stoeckli, F., Guillot, A., Slasli, A. M.,a, and Hugi-Cleary, D. (2002). Microporosity in carbon blacks. *Carbon 40*, issue 2, pp. 211-215, ISSN: 0008-6223

Stoeckli, F., Guillot, A., Slasli, A. M. & Hugi-Cleary, D. (2002). The comparison of experimental and calculated pore size distributions of activated carbons. *Carbon* Vol. 40, No. 3, pp. 383-388, ISSN: 0008-6223

Tomlinson, J. B., Freeman, J. J., Sing, K. S. W., and Theocaris, C. R. (1995). Rates of activation and scanning electron microscopy of polyaryllamide-derived chars. Carbon, Vol. 33, No. 6, pp. 789-793, ISSN: 0008-6223

Yuna, C. H., Parka, Y. H., Park, C. R.,. (2001). Effects of pre-carbonization on porosity development of activated carbons from rice straw, *Carbon 39*, pp.559–567, ISSN: 0008-6223

9

Study of Structure and Failure Mechanisms in ACA Interconnections Using SEM

Laura Frisk

Tampere University of Technology, Department of Electronics,
Finland

1. Introduction

The trends in the electronics industry have for several decades been for smaller size combined with greater functionality. One enabler for this trend has been development of new packaging solutions which has required the development of new materials and also new interconnections technologies. In the development of these technologies it has been essential to have effective tools to study the structure of the packages and their failure mechanisms. Due to the versatility of electronics packages concerning materials, structures, and functions a plethora of different methods have been used. These include for example electrical characterization technologies, x-ray, scanning electron microscopy (SEM), scanning acoustic microscopy (SAM), optical microscopy, differential scanning calorimetry (DCS), and thermomechanical analysis (TMA) (Chan et al., 2000; Jang et al., 2008; Yim & Paik, 2001).

This chapter concentrates on flip chip technology, which is one of the technologies developed to miniaturise an electronics package. In this technology a bare chip is attached directly onto a substrate without wiring needed to connect the chip. As the attachment is done with the active side of the chip towards the substrate, the chip is flipped before bonding. Hence the name flip chip. Flip chip technology has several advantages as it enables the production of small, very high density packages. In addition, it has good electrical performance due to the short interconnection path (Lau, 2000). This method can be used to attach a chip directly onto a substrate, but also as an attachment method in single-chip packages, such as a ball grid array (BGA) and a chip scale package (CSP), and in multi-chip modules.

One problem with this technology is that the quality of the joints is relatively difficult to assess. Semiconductor chips used in this technology may have hundreds of contacts which form an area array below the chip. In order to study the joints a cross-sectioning is often needed. Although cross-sections give lots of valuable information, they are restricted to very small area of a package and thereby often several cross-sections are needed. Additionally, the information of the interconnections may need to be increased using other techniques. For example scanning acoustic microscopy may be used to study the amount of delamination in a package. The cross-sections may be studied by optical microscopy. However, for detailed information scanning electron microscopy is the preferred method of analysis.

Currently mainstream flip chip technology is based on solder bumps. These can be produced using both traditional tin-lead and new lead-free solders. However, mounting environmental concern has increased interest in electrically conductive adhesives, as they are environmentally friendly. In addition to being lead free, they can be used with substrate materials which do not withstand soldering temperatures. Thus they can be used to solve the problem caused by the high reflow temperature needed by most lead-free solders (Li & Wong, 2006).

Compared to solders adhesive materials are more complex as they are polymeric materials containing conductive particles. They have several advantages which makes their use profitable. However, due to their complex structure quality of the interconnections made by these materials needs to be determined carefully to attain good reliability. There are two types of electrically conductive adhesives. In isotropic conductive adhesives (ICA) the concentration of the conductive particles is high and they conduct in all directions. On the other hand, in anisotropic conductive adhesives (ACA) the concentration of conductive particles is low and the adhesive conducts in z-direction only after the bonding process. This chapter will concentrate on ACA materials used in flip chip applications. This chapter will discuss specifically how SEM may be utilised to study the quality and failure mechanisms of ACA interconnections.

2. Polymeric interconnections for electronics

Electrically conductive adhesives used in electronics consist of polymer binder and conductive particles. Polymer resins used in these adhesives are inherently insulators. To obtain an electrically conductive adhesive they must therefore be filled with electrically conductive fillers, such as metal particles. In the following properties and materials of two main types of electrically conductive adhesives are discussed. Isotropic conductive adhesives (ICA) have high concentration of the particles and they conduct in every direction. These materials may be used to replace solders. If the concentration of conducting particles is low, an anisotropic conductive adhesive (ACA) is formed. ACAs conduct electrically only in a vertical direction and thereby may be used in very high density applications.

2.1 Isotropic conductive adhesives

Isotropically conductive adhesives are formed by adding enough conductive filler to a polymer matrix to transform it from an insulator into a conductor. This transformation has been explained by a percolation theory. When the concentration of conductive filler is increased, the resistivity of the adhesive drops dramatically above a critical concentration, and this is called the percolation threshold. It is believed that at this concentration the conductive particles contact each other forming a three dimensional network, which enables the conductivity. After the percolation threshold the resistivity decreases only slightly with increased concentration of the conductive filler. (Lau et al., 2003) The mechanical interconnection of an ICA joint is provided by the polymer matrix. If too high a concentration of the conductive filler is used, it may impair this interconnection. Thus the amount of conductive filler needs to be large enough to ensure good conductivity without sacrificing the mechanical properties of the adhesive (Lu, 2006). A typical volume fraction of the conductive filler is approximately 25 to 30 percent (Licari, 2005).

Several materials can be used in ICAs. The most widely used ICAs in the electronics industry are silver-filled epoxies, which also provide a high level of thermal conductivity. The popularity of epoxies is due to their excellent properties as a conductive adhesive. They have good adhesive strength, thermal stability and dielectric properties. Furthermore, they have good retention of these properties under thermomechanical stresses and under demanding conditions such as high humidity. However, other thermoset resins, such as silicones, cyanate esters, and cyanoacrylates, can also be used. Another option is thermoplastic resins. (Licari, 2005) Silver is the most commonly used filler material (Lau, 2003). The popularity of silver is due to its excellent conductivity and chemical stability (Morris 2005). Moreover, its oxide is highly conductive.

ICAs have been used in the electronics industry mainly as die-attach adhesives. However, lately they have also been proposed as an alternative to solders in surface mount and flip chip applications. For use in flip chip applications ICAs need to be carefully applied only on those areas which need to be conductive. Additionally, spreading of the adhesive should be prevented during the bonding process. A separate underfilling step is needed to improve the reliability of the joints. (Lau, 2003; Li, 2006) A typical cross section of an ICA flip chip joint is presented in Figure 1.

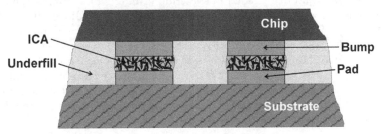

Fig. 1. Schematic illustration of an ICA flip chip joint with underfill.

2.2 Anisotropic conductive adhesives

In an anisotropic conductive adhesive (ACA) the concentration of conductive particles is below the percolation threshold (Lau, 1995; Licari, 2005) and the adhesive does not conduct before the interconnection is formed. Typically the number of particles is 0.5% - 5 % by volume (Licari, 2005) but depends largely on the size and shape of the conducting particles and on the application the ACA is used in (Watanabe, 2004). Normally the particles are randomly dispersed in the matrix, but adhesives having uniformly dispersed particles have also been developed (Ishibashi & Kimura, 1996; Jin et al., 1993; Sungwook & Chappell, 2010).

During the ACA attachment process the adhesive is placed between the mating contacts. The ACA interconnection is established by applying pressure and heat simultaneously to the interconnection. When the temperature is raised the adhesive matrix will transform into low viscosity fluid (Tan et al., 2004), which allows excess adhesive to flow from the joints and fill the spaces around the contacts forming a physical connection between the parts to be attached. The conductive particles are trapped between the contacts and deform forming an electrical connection. As a result, electrical conduction is restricted to the z-direction and the electrical insulation in x-y directions is maintained. During cooling residual stresses are formed as a result of contraction of the adhesive matrix. In addition, residual stresses form when the adhesive shrinks during curing. However, it has been shown that the residual stresses formed during cooling dominate (Kwon & Paik, 2004). This contraction builds up a sufficient force to create a stable, low-resistance connection. A typical cross section of an ACA flip chip joint is shown in Figure 2.

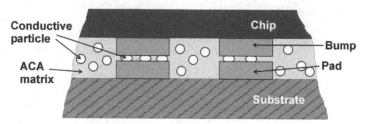

Fig. 2. Schematic illustration of an ACA joint between a chip and a substrate.

ACA joints have several advantages compared to underfilled solder interconnections. As the ACA process is solderless, there is no lead or alpha emission (Zhong, 2005). Moreover, the process is fluxless and no cleaning is required (Zhong, 2005). Furthermore, the process temperature is lower than that needed in soldering (Yim & Paik, 1998), which enables the use of heat sensitive or non-solderable substrate materials (Uddin et al., 2004). As the polymer matrix protects the contacts from mechanical damage and no underfilling is required (Lai & Liu, 1996), the ACA process costs less due to fewer processing steps (Yim & Paik, 2001). The ACA joining also enables very high interconnection density. On the other hand, the ACA joint has higher contact resistance and lower current capability than that made with solder (Jim & Paik, 1998; Zhong, 2005). Since the ACA has no self-alignment capability, a special bonding machine is needed for accurate alignment. During the bonding process heat and pressure also need to be applied simultaneously.

2.2.1 Materials used in ACAs

Both thermoplastic and thermoset materials and their mixtures have been used as an ACA matrix. Initially, the ACAs were made of thermoplastic materials, as they have better reworkability and pot life (Lau, 1995). However, their stability at high temperatures is not good and the thermoplastic material is not strong enough to hold the conducting particles in position, which increases the contact resistance of the joint (Asai et al., 1995; Kim et al., 2004). Thermoset adhesives were developed to overcome the problems with thermoplastic adhesives. Thermoset adhesives are stable at high temperatures and enable low joint resistance. Epoxies are commonly used as an ACA matrix due to their good properties. Epoxies have excellent adhesion to a variety of substrates, due to the highly polar hydroxyl and ether groups (Luo, 2002). In addition, they have high glass transition temperature (T_g) and favourable melt viscosity (Kim et al., 2004; Yim & Paik, 1999). Furthermore, epoxies give low contact resistance and by selecting suitable curing agent long self-life and fast-cure properties can be achieved. The epoxy resin forms a crosslinked structure during bonding with good mechanical properties. However, their reworkability is problematic, as they are not thermally reversible and do not dissolve in common organic solvents (Lau, 1995).

The electrical conduction in ACA is formed by the conductive particles. The size, concentration and material of the conductive particles depend on the application area and on the manufacturer. Typically the conductive particles are approximately 3-10 um in size. Nowadays, the most common conductive particles are nickel, which may be gold plated, and metal plated polymer particles. However, other materials, such as carbon fibres and solder balls, have also been used (Asai et al., 1995). The polymer particles are made of polystyrene cross-linked with divinyl benzene (Asai et al., 1995) and the metal plating on them may be of nickel, silver, or gold (Liu, 1996). The polymer particles are pliant and during the bonding process they deform, thereby forming the connection to the contacts. The deformation of the rigid nickel particles during the bonding process is less than that of the soft particles and the contact area formed is smaller. However, if the bonded contacts are made of softer metal, such as gold or copper, these contacts deform during bonding increasing the contact area with the rigid particles (Divigalpitiya & Hogerton, 2004; Yim & Paik, 1998; Frisk & Ristolainen, 2005; Frisk & Kokko, 2006).

In the flip chip process the interconnection is formed between pads on the substrate and bumps on the chip. However, as the cost of bumping may be unattractive in certain applications, bumpless chips are also used. The most commonly used bump materials are gold plated nickel, and gold. The gold bumps can be manufactured using an electroplating process. Copper bumps formed by similar electroplating techniques have also been considered as an alternative to gold because of their lower cost (Lau, 2000, Lau, 2003). However, copper oxidizes and corrodes easily, which may cause problems if it is used without plating. The nickel bumps can be made using an electroless plating process. This process has high potential for cost reduction, as it enables metal deposition directly on the aluminium pads on the chips. Thus the costly equipment needed in the electroplating process for sputtering, photoresist imaging, and electroplating is eliminated. The gold bumps may also be processed using a modified wire bonder to form stud bumps. The advantage of this process is that bumps can be formed on single chips in addition to whole wafers. (Lau, 1995)

2.2.2 The ACA bonding process

ACAs can be used either as films (ACF) or as pastes (ACA or ACP). The type of the ACA affects the bonding process and the equipment needed. ACFs are typically supplied in a reel and a dedicated in-line bonding machine is needed for cutting, aligning, and tacking to achieve high assembly speed. On the other hand, ACPs can be applied either by printing or by dispensing using a syringe. Even though the ACF process requires special equipment, ACFs are often used as they offer advantages compared to ACPs. The ACF process consumes less material than the ACP process. Moreover, the ACP process may destroy the randomness of the particle distribution leading to problems in process quality.

In the ACF bonding process the adhesive film is first cut to the correct size to cover the bonding area. After this the adhesive is aligned to the substrate and pretacked using light pressure and low temperature to attach the ACF to the substrate. After pretacking the carrier film on the ACF is removed. Next a chip is picked by a flip chip bonder. Typically a special flip chip bonder capable of simultaneously applying pressure and heat is used. The bumps on the chip and the pads on the substrate are aligned. The chip is pressed onto the substrate and heat is applied to the chip and the polymer matrix is cured. A schematic illustration of the bonding process used is presented in Figure 3. In case ACP material is used the steps a and b are replaced by deposition of the ACP. After this steps c and d are performed similarly to the ACF process.

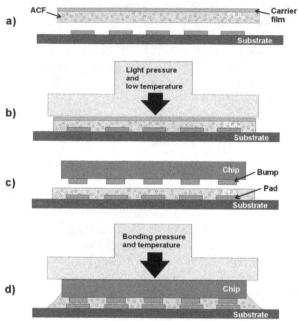

Fig. 3. Schematic illustration of the ACF flip chip bonding process: a) placement of ACF on the substrate, b) prebonding, c) alignment of the chip and the substrate and d) final bonding.

3. Evaluation of the quality of ACA interconnections

The quality of the ACA interconnections may be studied using several techniques. In general, a good quality ACA interconnection is characterized by low contact resistance and good mechanical properties. Therefore, electrical measurements may be done to assess the quality. However, often this is not possible due to the design of the semiconductor chip. Electrical measurements may show alignment and planarity problems as higher resistance values. However, this is not always the case, as it is possible that the electrical connection seems good even though there are problems in the joints causing reliability problems during use. The alignment of an interconnection may be studied through the substrate in the case of transparent substrates such as glass or thin polyimide film. However, often the most effective way to examine both alignment and the structure of the interconnection is to make a cross-section of the structure and study them using either optical or scanning electron microscope. SEM especially is often very a effective tool yielding a plethora of information of the joint, which is important for optimisation of the bonding process.

The mechanical properties of the ACA interconnections may be studied using adhesion testing, which will indicate how well the ACA material is attached to the substrate and the chip. In this technique both adhesion strength and failure mechanism during testing will give valuable information. In the following several different parameters affecting the quality of ACA interconnections are discussed. Special attention is paid to the information obtained from the interconnections using SEM analysis.

3.1 SEM analysis of ACA interconnections

As mentioned above, both optical and scanning electron microscopy may be used to study the cross-sections of ACA interconnections. Although, studying the cross-sections is typically very effective it has some drawbacks. The number of particles in ACA interconnections is typically quite low. Thus, when a cross-section of an interconnection is studied, the probability of seeing particles in interconnections is small even if there are sufficient particles in the interconnections to ensure proper joining. From a cross-section only one side of a chip is seen. Therefore, to determine planarity and alignment issues several cross-sections are needed. Furthermore, making a cross-section of a sample destroys it thereby gravely restricting further analysis. Optical microscopy may give valuable information especially when planarity and alignment are concerned. However, typically it does not give very good detailed information. Especially, if interfaces of different materials are studied an optical microscope does not give reliable information. On the other hand, SEM is often a very powerful tool to determine the quality and structure of an ACA interconnection. However, SEM analysis benefits from information of other analysis methods if they are available such as, for example, electrical characterisation and scanning acoustic microscopy.

For SEM analysis the quality of the cross-sections needs to be good and they need to be clean. Typically epoxy moulding is used followed by grinding and polishing. However, other materials are possible such as acrylics. Use of high temperature mould materials may cause problems depending on the structure studied and its materials. As the ACA structure has many different materials the analysis is often challenging and the parameters used for SEM need to be determined according to the samples studied. Additionally, the area of interest in the interconnection affects the parameters. Both thin gold and carbon layers may be used to make samples electrically conductive. However, gold gives better quality of analysis and is recommended if elemental analysis is not needed.

3.2 Bonding parameters

Bonding parameters are the key factors when the quality and the reliability of the ACA interconnections are considered. The ACA process makes the bonding parameters especially important as the complex mechanical, rheological, and chemical properties of the ACA materials need to be considered (Dou, 2006). The most important parameters in the ACA bonding process are time, temperature and pressure. However, other bonding parameters, such as application rates of pressure and temperature, also affect the quality and reliability of the joints (Ogunjimi et al., 1996; Whalley et al., 1997). Moreover, the parameters may interact with each other.

Finding optimum process parameters necessitates careful study. Quite often the quality of the bonding parameters cannot be determined on the basis of electrical or adhesion measurements only and cross-sections are needed for verification of the interconnection quality. For thermoset adhesives the bonding temperature together with the bonding time determine the degree of cure of the adhesive matrix. The higher the bonding temperature used the higher the degree of cure of the adhesive matrix is within the same amount of time (Chan & Luk, 2002a; Rizvi et al., 2005; Tan et al., 2004; Wu et al., 1997) as the higher temperature accelerates the crosslinking reaction (Uddin et al., 2004). Similarly, longer bonding time increases the degree of cure. Both bonding time and temperature needs to be high enough to achieve adequate curing of the adhesive matrix as the mechanical and chemical properties of the ACA have been found to depend heavily on the degree of cure (Wu et al., 1997). Too low degree of cure is often seen as high contact resistance values and also as inadequate adhesive strength.

The bonding pressure determines the deformation of the conductive particles. If the bonding pressure is too low the conductive particles cannot make good contact with the bonded surfaces and the contact resistance will be high (Chan & Luk, 2002b; Lau, 1995). In addition, the reliability of this kind of joint is poor (Frisk & Ristolainen, 2005; Lai & Liu, 1997). Examples of only slightly deformed particles are shown in Figure 4. In this case the variation is assumed to be caused by too fast curing of the adhesive during the bonding process. This adhesive has been designed to cure very quickly. Consequently it may have started to cure before the adhesive had flowed properly, leaving the particles insufficiently deformed.

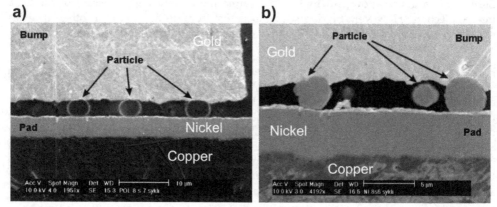

Fig. 4. An example of slightly deformed particles: a) gold coated polymer particles and b) nickel particles.

When the bonding pressure is increased, the contact resistance typically decreases sharply at first before evening out (Yim & Paik, 1998; Yin et al., 2003). This is caused by greater deformation of the conductive particles leading to a larger contact area between the particle and contacts (Kwon et al., 2006; Yin et al., 2003). Figure 5 shows examples of properly deformed particles. However, if the pressure is increased too much, the contact resistance may start to increase again (Chan & Luk, 2002b; Yim & Paik, 1998). If metal plated polymer particles are used, too high pressure may crush the particles (Wang et al., 1998) and lead to direct contact between the bump and the pad. The cracking of the metal plating may separate it from the polymer core and reduce the amount of conductive path between the pad and the bump (Yim & Paik, 1998). Another problem with too high bonding pressure is elastic stress formed in the chip or in the substrate (Frisk et al., 2010; Lai & Liu, 1996). In some cases, the high temperature used during the bonding process may soften the substrate material used and it will deform markedly during the bonding process.

a) b)

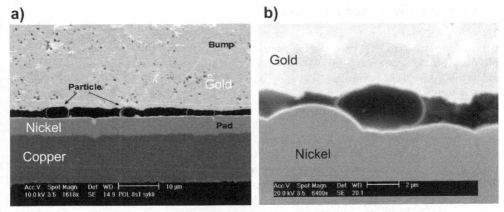

Fig. 5. a) and b) examples of marked deformation of gold plated polymer particles.

If rigid particles are used, the bump and the pad material also have a strong effect on the bonding pressure. With soft bump materials, such as gold or copper, the particles will sink into the bump forming a strong bond. If high bonding pressure is used, the particles will sink completely into the bumps and direct contact between the pad and the bump will be formed. An example of direct contact is presented in Figure 6. The hard nickel-plating on the pads prevented the penetration of the particles into the pads.

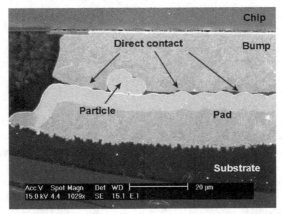

Fig. 6. Penetration of the nickel particles into the copper bump and direct contact between the bump and the pad.

3.3 Effect of the substrate material

Poor substrate and chip quality may cause coplanarity problems. This may increase the contact resistance, as not all joints have adequate deformation of the particles. For the flip chip process to be usable with the organic substrates, the planarity of the substrate is very important. Planarity issues often need to be determined using cross-sections. In the following examples of SEM analysis used for quality studies of ACA interconnections with different substrate materials are given.

3.3.1 Glass fibre reinforced substrates

Glass fibre reinforced materials are commonly used as substrates in electronics. The most widely used material is FR-4, which is a grade designated by the National Electrical Manufacturers Association (NEMA), which determines that the material is flame resistant, is primarily epoxy based, and has woven glass fibre reinforcement (Coombs, 2000). As the resin material between different FR-4s may vary, the properties of FR-4 materials are not identical. For example typical T_g of FR-4 is between 130 and 140 °C or between 170 and 180 °C. The popularity of the FR-4 is based on its good properties, availability, and low cost. When rigid glass fibre reinforced substrates are used in ACA applications, coplanarity problems may arise due to the woven structure of the substrate (Frisk & Cumini, 2009). During the bonding process the high temperature may soften the resin, which leads to its deformation under the pads. This deformation has been found to depend on the orientation of the glass fibres in the substrate and affect the electrical conductivity and reliability of the joints (Liu et al., 1999).

Deformation of substrate in ACA interconnections was studied using cross-sections and SEM (Frisk & Kokko, 2006; Frisk & Cumini, 2009; Frisk et al., 2010). SEM has proven to be a very effective method for such studies as the different materials and their interfaces may be clearly seen. Figure 7 shows micrographs of a FR-4 substrates after an ACA bonding process. Deformation of the substrates may be seen between contacts. Deformation is especially considerable in the areas where the glass fibres were far from the surface. This varying deformation of the substrate causes pressure variation in the joints (Pinardi, 1998). The varying pressure is important as it may result in different deformation of the particles leading to variation in the contact resistance and also impairing the reliability of the joints.

Quite often in flip chip applications with ACA materials very high wiring densities are needed. These are difficult to achieve with the traditional FR-4 substrates shown in Figure 7 One possibility to meet these demands is to use sequential build-up (SBU) processes. However, in this process conductor and dielectric layers are formed one after another on a rigid core board, which may be an FR-4 glass reinforced laminate (Tagagi et al., 2003). The electrical connection between the core board and the build-up layers is formed using microvia technologies. An example of a substrate made with the SBU process is presented in Figure 8. The typical dielectric materials used in SBU build-up layers are resin-coated copper foil (RCC or RCF), thermally cured resin, and photo-imageable resin (Tagagi et al., 2003). The most widely used dielectric material in the SBU process is RCC. The RCC is formed by adding a layer of resin to a thin copper foil, which is laminated to the core board. A typical resin material is epoxy. The RCC has several advantages and is suitable for processing in standard printed circuit board processes.

As the RCC layer does not have glass fibres, it is more pliable than an FR-4 substrate. Its effect on the interconnection structures was studied using cross-sections and SEM (Frisk & Kokko, 2006). During the bonding process the depression of the copper pads into the RCC was found to be much stronger than the depression into the FR-4 substrate. In Figure 9 an example of the depression is presented for test samples with the RCC test substrate. As can be seen, the RCC has deformed markedly more during the bonding process than the pure FR-4 substrates shown in Figure 7. Furthermore, some deformation of the FR-4 substrate beneath the RCC has occurred. Although the deformation of the RCC is greater, it is almost identical under every pad leading to more uniform distribution of pressure. This causes the deformation of the particles to be identical in the joints and increases reliability compared to the substrate without the RCC.

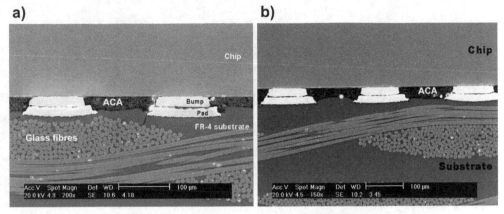

Fig. 7. a) and b) micrographs showing deformation of a FR-4 substrate

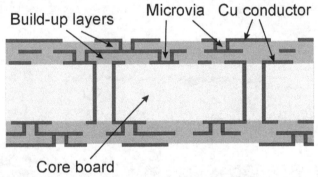

Fig. 8. Schematic cross-section of a printed circuit board made with SBU process.

Deformation of the particles between these test boards was also studied with SEM. With the test substrate, which did not have the RCC, the particles sank into the copper bump. However, the fairly thick nickel plating on the pads prevented the particles from sinking into it and caused deformation of the particles, as can be seen in Figure 10 a). In addition, due to relatively high magnification of the SEM micrograph the gold layers on both conducting particles and pad can be easily seen. The thinner nickel plating on the RCC test board enabled the particles to sink into the pad, as can be seen in Figure 10 b). Moreover, as the RCC gives in more under the pads during bonding than the FR-4 substrate, the deformation of the particles was less on the substrates with the RCC. Deformation of the particles is important for the reliability of the joints. Sufficiently deformed particles form a strong atomic interaction between the pad and the bump creating increased stability of the joint (Lai & Liu, 1996). Using SEM the interface between the particles and the pad or the bump may be studied and problems such as thin layers of polymer matrix may be detected. In both Figures 10 a) and 10 b) a good contact of particles is seen to both the pad and the bump .

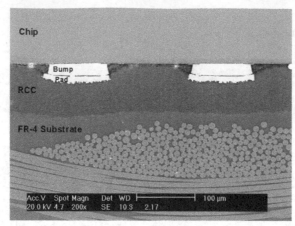

Fig. 9. Micrograph presenting the immersion of the pads in the RCC, when high bonding pressure is used.

a)   **b)**

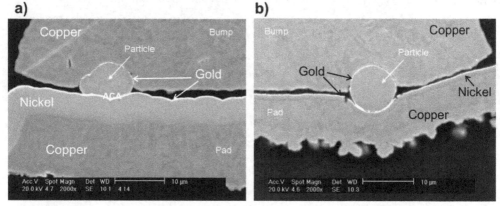

Fig. 10. a) Micrograph presenting the deformation of the rigid nickel particle, when a FR-4 substrate is used. b) Micrograph presenting the immersion of the rigid nickel particle in the pad and the bump, when the substrate with the RCC is used.

3.3.2 Flexible substrates

ACA materials are often used with flexible substrates, which are fabricated using pliable unreinforced polymeric materials. Flexible substrates have several advantages compared to fibre-reinforced substrates and lighter and thinner products can be produced using them. Flexible substrate may absorb stress, which may be important for the reliability of the interconnections, especially in flip chip applications. In addition, the thermal transfer through a thin substrate is more effective. Furthermore, very high density substrates are available with flexible substrates and this is often critical for ACA applications. On the other hand, the thinness of the substrate may cause problems in the stability of the construction. Moreover, the cost of the flexible substrates is higher than that of the rigid substrates. (Coombs, 2001)

The pliability of the flexible substrates may cause some problems during the bonding process. Figure 11 shows SEM micrographs of ACA interconnections with flexible liquid crystal polymer (LCP) substrates with two different pressures. A marked deformation of the liquid crystal polymer film with high pressure may be seen. Such deformation may cause problems. On the other hand, it may also even out some planarity problems as deformation can absorb the height variations (Connell, 1997; Savolainen, 2004) and thereby increase the quality of the interconnection. However, in some cases deformation may cause cracking of the wiring and thereby lead in to reliability problems. With the lower pressure the pressure exerted to the particles may not be high enough and therefore cause reliability problems.

a)   b)

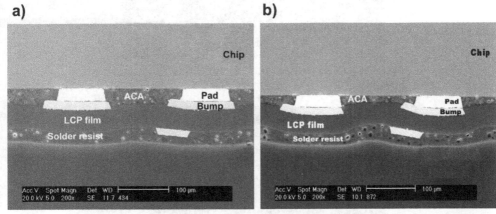

Fig. 11. An example of deformation of the LCP substrate when a) low bonding pressure and b) high bonding pressure was used.

With thin flexible substrates routing may also be critical to the distribution of the pressure, if the substrate has several conductive layers (Lai & Liu, 1996). If double sided flexible substrates are used they may have wiring on both sides of the bonding area. This may cause uneven distribution of the pressure and deformation of contact areas. Figure 12 shows an interconnection with a polyimide substrate having double sided wiring. As can be seen, the gold bump has markedly deformed during the bonding process because of the wiring on the other side of the substrate. A similar effect may be seen in the LCP substrates in Figure 11. However, as the solder resist on the LCP substrate evened out the effect of the wiring, there is clearly less deformation. Consequently, the design of flexible circuitry is very important for good quality interconnections. Furthermore, such quality problems are difficult to detect without cross-sectioning.

Another problem in a substrate may be overetching of the pads. As flexible substrates are often used in application where very high density substrates are needed such as attachment of driver chips is display application, overetching may cause marked problems. It reduces the contact area and may cause alignment and planarity problems. Figure 13 shows a SEM micrograph of a interconnection with an overetched polyimide (PI) substrate. In the original substrate layout the pad width was designed to be slightly greater than the bump. However, due to overetching the size of the pad is clearly less than that of the bump. Such overetching may be seen when substrates are quality checked before use. However, cross-section is a good way to evaluate the effect of overetching on an interconnect.

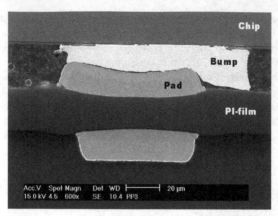

Fig. 12. Effect of the tracks on joint deformation on the PI substrate when the tracks are on both sides of the substrate.

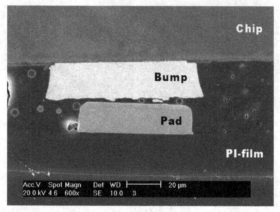

Fig. 13. An overetched pad on the PI substrate.

4. Failure mechanisms of the ACA flip chip joints

In general, a good ACA joint is characterized by low contact resistance. A key issue for the ACA to function properly is the retention of this contact resistance during the operational life of the product. Failure of a product can be defined as its inability to perform its intended function [For an ACA flip chip joint this typically means too great an increase in contact resistance. There is no single specific definition of failure of ACA joint resistance, as this depends largely on the application.

To improve the reliability of interconnections the reasons for their failure and the failure mechanisms must be understood. During the design phase reliability is typically assessed using accelerated environmental testing. The aim of such testing is to predict the future performance of a product in a shorter period of time than the service life of the product. Accelerated life tests can also be used to detect failure mechanisms occurring in products under different conditions of use. The acceleration is accomplished by using elevated stress levels or higher stress cycle frequency during testing compared to those under normal operational conditions of the product (Suhir, 2002). Depending on the condition failures may occur through several mechanisms. It is important that the testing conditions are determined so that the failure mechanisms during testing are similar to those occurring under normal conditions of use. The test conditions depend decisively on the application of the product. For example, the test conditions for products used in military and space applications are much more rigorous than those for consumer electronics.

Several different accelerated life tests have been used to study the reliability of ACA joints; see for example (Frisk & Ristolainen, 2005; Frisk & Cumini, 2006; Jang et al., 2008; Kim et al., 2004; Lai & Liu, 1996; Saarinen et al., 2011). These include high temperature aging tests, temperature cycling tests, high temperature and high humidity tests, and humidity and temperature cycling tests. The reliability and failure mechanisms of ACA interconnections depend on several factors, including the materials and bonding parameters. The materials used in the pads, bumps, and conductive particles need to be compatible with each other. The substrate material may also have a marked influence on the reliability of the joints. As the properties of the joints are much influenced by the bonding process, it is important that optimum bonding parameters are determined and used in the bonding process. In the following failure mechanisms and analysis of ACA interconnections are discussed.

4.1 Failures of ACA interconnections with rigid FR-4 substrates under thermal cycling

One of the major problems in ACA assembly using organic substrates is the great difference between the coefficient of thermal expansions (CTE) of chip and substrate. The bonding process is done at elevated temperatures. When the package is cooled down the contraction of the silicon chip is very small due to its low coefficient of thermal expansion and high Young's modulus (Kwon et al., 2005). On the other hand, the contraction of the substrate is much greater. At low temperatures this causes stresses to form between the chip and the substrate and the ACA flip chip package to warp downwards. At temperatures below its T_g the adhesive matrix holds the chip and the substrate together enabling this warpage (Kwon & Paik, 2006). However, when the T_g of the adhesive is exceeded, its mechanical strength decreases and it cannot provide mechanical support between the substrate and the chip. The warpage is evened out and both chip and substrate may expand with their inherent CTEs (Kwon & Paik, 2006). The warpage of the ACA package is presented in Figure 14. If the ambient temperature fluctuates, as in a temperature cycling test, the flip chip package warps repeatedly. The warping reduces the shear stress in the joints and a greater degree of warp has been reported to decrease the shear strain in the joints (Kwon et al., 2005).

The amount of warpage depends on the difference between the coefficient of thermal expansions of chip and substrate, but also on their stiffness. When a rigid fibre reinforced substrate is used, the shear stress caused by CTE mismatches is localised between the pad and the bump as the of deformation the substrate is less than that of the adhesive matrix (Lai et al., 1998). On the other hand, deformation of an unreinforced flexible substrate occurs much more easily due to its low modulus, and some of the shear stress may be absorbed by the substrate (Connell et al., 1997). This decreases the shear stress in the interconnection reducing the delamination and increasing the reliability.

Above T$_g$ of ACA

The elastic modulus of the ACA is low at temperatures above its T$_g$ allowing the chip and the organic substrate to expand freely. No warpage occurs.

Below T$_g$ of ACA

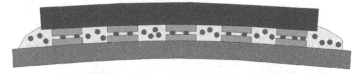

At low temperatures the organic substrate contracts more than the chip and the package warps downward.

Fig. 14. Warpage of the ACA flip chip package during temperature cycling test.

ACA interconnections with rigid FR-4 substrate were exposed to thermal cycling between -40 °C and +125 °C (Frisk & Kokko, 2006). When the failed test samples were studied using SEM, delamination was found in several test samples after testing. This delamination is probably caused by shear stress between the pad and the bump formed due to differences in the coefficient of thermal expansion of the substrate and the chip. As shown in Figures 16 a), 16 b), and 16 c) with SEM the delamination is seen clearly and its place can be determined. In some cases delamination may be less pronounced and therefore harder to detect. In Figure 16 c) another example of delamination which is more difficult to detect is shown.

Another phenomenon seen in these samples was cracking of the substrate material. This is assumed to be caused by repeated warping of the substrate as the duration of testing was several thousand cycles. This type of cracking was typically found in areas where the glass fibres were far from the surface of the substrate. They started from the corners of the pads and often continued into the substrate until they reached the glass fibres. Examples of such cracking are presented in Figure 16. The formation of such cracks has probably facilitated delamination between pad and the bumb by providing sites for crack initiation, as the cracks often connected to the delamination between the pad and the bump. An example of this is presented in Figures 16 b) and 16 c).

It has been suggested that during thermal testing above the T$_g$ of the ACA matrix sliding between the pad and the bump occurs (Kwon et al, 2006; Uddin et al., 2004). This may break the conductive particles and lead to failure. Furthermore, this phenomenon may cause fretting and thereby the formation of oxides on the conductive surfaces, which increases the resistance of the joints above an acceptable value. When ACA interconnections having nickel particles or gold coated polymer particles, were studied after thermal cycling clear indication of this phenomenon was seen. Figure 17 a) shows a SEM micrograph of polymer particles form a failed ACA interconnection. In the picture cracking of the gold layer in the particle can be clearly seen. Additionally, agglomeration of material under the particles can be seen indicating failure caused by this sliding. Figure 17 b) shows a similar situation for nickel particle and similar agglomeration of material.

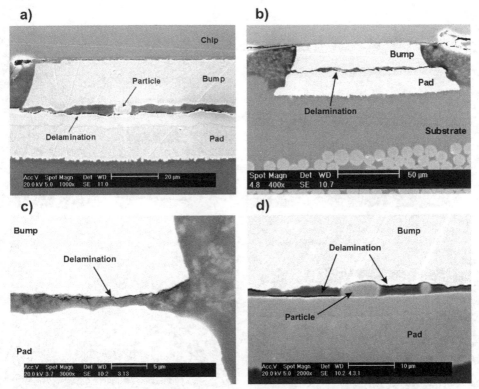

Fig. 15. Examples of delamination in test samples with FR-4 substrates after thermal cycling a) clear delamination between ACA and pad, b) delamination between bump and pad which continues to ACA-chip interface, c) less pronounced delamination, and d) delamination varying between the ACA-pad interface and the ACA-bump interface.

4.2 Failures of ACA interconnections with rigid FR-4 substrates under humidity testing

In addition to temperature changes, humidity has been found to have a major influence on the reliability of ACA interconnections. Under humid conditions the adhesive matrix may deform as it absorbs water. The adhesive matrix may also relax due to the increased temperature. The effect of water absorption depends on the structure of the adhesive and also on the duration of the exposure. The most common adhesive material is epoxy. It has been suggested that the water absorbed in the epoxy polymer has two states to which the water molecules can diffuse. The water may either fill the free volume in the polymer matrix or form hydrogen bonds with the epoxy polymer. If the water is hydrogen-bonded, it causes swelling of the adhesive matrix (Chiang & Fernandez-Garcia, 2002; Luo et al., 2002). This swelling due to humidity typically increases with temperature, but decreases sharply across the T_g of the polymer (Wong et al., 2000). Swelling due to moisture differs significantly between the materials used in flip chip packages causing the formation of hygroscopic stresses in the structure (Mercado et al., 2003; Wong et al., 2000).

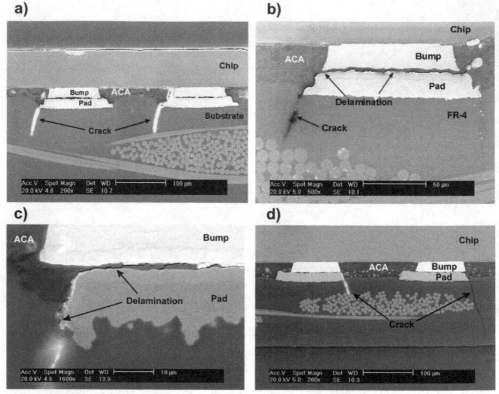

Fig. 16. Examples of cracks in FR-4 substrates: a) marked deformation and clear cracking of the epoxy matrix, b) cracking which continues between the pad and the bump, c) less pronounced cracking which continues between the pad and the bump, and d) cracking of thin FR-4 substrate

The swelling of the adhesive matrix may be marked (Mercado et al., 2003) and concurrent with thermal expansion may cause the conductive particles to lose contact. However, it has been reported that the absorbed water may also weaken the mechanical properties of the adhesive. Unlike the hydrogen-bonded water, the water filling the free volume in the epoxy polymer does not cause swelling, as it occupies a volume that already exists (Chiang & Fernandez-Garcia, 2002). However, it acts as a plasticizer affecting the mobility of the chains and increasing chain flexibility, which decreases the T_g of the polymer. As the water acts like a plasticizer it may also impair the mechanical strength of the adhesive.

The substrate material may also have a great influence on the reliability of the joints under humid conditions (Frisk & Cumini, 2006). If the substrate material absorbs water, it penetrates the interfaces more easily and may cause delamination. A large amount of moisture in the substrate also facilitates the moisture absorption of the adhesive matrix, and may accelerate the formation of moisture related failures. The effect of humidity at elevated temperatures on the ACA joints was studied using an 85°C/85RH test. Flexible polyimide was used as a substrate material. When studied after testing using SEM, every test sample with the PI substrate which showed an open interconnection after testing also showed delamination. An example of the delamination is presented in Figure 18. The moisture absorption of polyimide is marked and it is assumed to be the reason for the formation of delamination during testing.

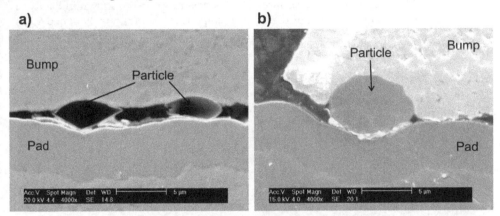

Fig. 17. Micrograph of a failed ACA interconnection after thermal cycling a) with polymer particles and b) with nickel particles.

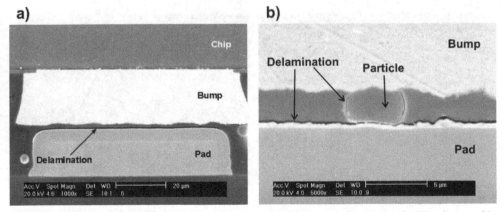

Fig. 18. a) An example of delamination after constant humidity testing on the PI substrate.b) close view of delamination after constant humidity testing on the PI substrate and on particle.

ACA technique was also studied with thinned silicon chips. When silicon chips are thinned below 100 um, they become pliant and can be used in solutions where they are bent. Consequently, they can be used in flexible electronics. Thinning also allows the chips to dissipate more heat, which is important when the densities of the packages increase.

However, thinned chips have certain drawbacks. They are more fragile than thicker chips, which needs to be taken into account when thinned chips are handled. Special tools may be needed since the fragile edges of the thinned chips are easily broken during handling. During the thinning and dicing processes a considerable amount of stress may be induced in the chips. It has been found that the thinning of the chips changes the shear stress distribution in a package (Frisk et al., 2011). In the analysis 50 μm thick chips were used instead of approximately 500 μm thick chips used in other studies. Very strong delamination was seen in these test samples which indicates failure mechanisms which is different from the one seen with the thicker chips. Examples of this delamination are shown in Figures 19 a) and 19 b). Furthermore cracking of the thinned chips was seen (Figure 19 c) and 19 d)). However, most of this occurred during the bonding process already before testing.

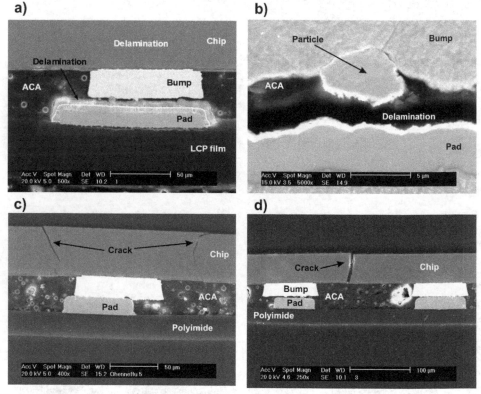

Fig. 19. Micrographs for ACA interconnections with thinned chips: a) marked delamination after humidity testing, b) close view of delamination under particle, c)and d) cracks in the thin chips.

5. Conclusion

Anisotropic conductive adhesives (ACA) are an interesting interconnection method for several applications. Due to the low cost of the ACA process and capability for high density they nowadays dominate many fields for instance attachment of chips in radiofrequency identification (RFID) tags and attachment of driver chips in display applications. ACA interconnections often show the typical trend in electronics for very small size with high functionality. In general this means high density of contacts in a chip and often also a large number of contacts per chip. Such applications are often very challenging both for studies of the interconnection process and the quality and reliability of the interconnections. Making cross-sections of the interconnections has proven to be efficient way to obtain detailed information about the interconnections structures, their quality and failure mechanisms, and this has been used effectively with other characterisation methods such as scanning acoustic microscope, DSC, and x-ray. Due to the small features currently common in electronics applications SEM is often the preferred method for examinations compared to optical microscopy.

In this chapter several SEM analyses of ACA interconnections were described. Many of these have been critical to both understanding the bonding process of these materials and also for the development of the ACA interconnection techniques and their reliability. Studying cross sections with SEM has been shown to be an effective way to analyse several of the failure mechanisms found in ACA interconnections. However, in general a good understanding of SEM analysis technique is needed for analysis due to the complexity of the ACA structure. ACA interconnection has many interfaces and the failure may occur on any of these or in the bulk materials of the structure. For effective analysis the critical parts need to be already understood when the analysis is made, and therefore, a good understanding of the technique and materials is needed. Lately the use of ACA technology has increased and it has been adopted on new areas for example high temperature electronics and sensor applications. In the future, this will increase the need for detailed knowledge of this interconnection technique. Therefore, it is critical that studies such as presented in this chapter are continued as they give vital information for both development and applicability of this technique.

Making cross-sections of studied samples is favoured in ACA flip chip applications due to the difficulties of studying the interconnections below a component. This also applies to many other interconnections and packaging technologies and similar methods have been used successfully in other applications. For example in the research of lead free solders cross-sections are systematically used for failure analysis and studies related to the microstructure of the interconnections. Additionally, in other techniques, such as flex on board attachments for example, in which flexible substrate is attached to a rigid substrate, and small packaging solutions such as Chip Scale Package (CSP) cross-sectioning and SEM analysis is often needed to determine the structure. In the future, the size of interconnections in electronics will decrease and their number will increase. As a consequence, the need for techniques capable for the analysis of such structures will increase markedly. SEM has proven to be an extremely useful tool for analysing electronics structures as it is relatively fast, typically easily available, and capable for analysing small features.

6. Acknowledgment

I would like to thank my colleagues Dr. Kati Kokko, M.Sc. Kirsi Saarinen, M.Sc. Janne Kiilunen, and M.Sc. Sanna Lahokallio, and, additionally, my former colleague Dr. Anne Cumini for their help in this research. Furthermore, I would like to thank the staff at the Institute of Material Science at Tampere University of Technology.

7. References

Asai, S., Saruta, U., Tobita, M., Takano, M., and Miyashita, Y. (1995) Development of an Anisotropic Conductive Adhesive Film (ACAF) from Epoxy Resins. *Journal of Applied Polymer Science*, Vol. 56, No. 7, 1995, pp. 769-77. ISSN 1097-4628.

Chan, Y., Hung, K., Tang, C., and Wu, C. (2000) Degradation Mechanisms of Anisotropic Conductive Adhesive Joints for Flip Chip on Flex applications. *Proceedings of 4th IEEE Conference on Adhesive Joining and Coating Technology in Electronics Manufacturing*, ISBN 0780364600, Espoo, Finland, June, 2000.

Chan, Y. C. and Luk, D. Y. (2002a) Effects of Bonding Parameters on the Reliability Performance of Anisotropic Conductive Adhesive Interconnects for Flip-chip-on-flex Packages Assembly I. Different Bonding Temperature. *Microelectronics Reliability*, Vol. 42, No. 8, 2002, pp.1185-94. ISSN 0026-2714

Chan, Y. C. and Luk, D. Y. (2002b) Effects of Bonding Parameters on the Reliability Performance of Anisotropic Conductive Adhesive Interconnects for Flip-chip-on-flex Packages Assembly II. Different Bonding Pressure. *Microelectronics Reliability*, Vol. 42, No. 8, 2002, pp.1195-1204. ISSN 0026-2714

Chiang, M. and Fernandez-Garcia, M. (2002) Relation of swelling and T_g depression to the apparent free volume of a particle-filled, epoxy-based adhesive. *Journal of Applied Polymer Science*, Vol. 87, No. 9, 2002, pp. 1436 - 44. ISSN 1097-4628

Connell, G., Zenner, R., and Gerber, J. (1997). Conductive Adhesive Flip Chip Bonding for Bumped and Unbumped Die. *Proceedings of the 47th IEEE Electronic Components and Technology Conference*, ISBN: 0-7803-3857-X, San Jose, CA, USA, May, 1997.

Coombs, C. (2001). *Coombs' Printed Circuits Handbook*. McGraw-Hill Professional, 5th edition, ISBN 0071350160.

Divigalpitiya, R. and Hogerton, P. (2004). Contact Resistance of Anisotropic Conductive Adhesives. *Journal of Microelectronics and Electronic Packaging*, Vol. 1, No. 3, 2004, pp. 194-9. ISSN 1551-4897.

Frisk, L. and Ristolainen, E. (2005) .Flip Chip Attachment on Flexible LCP-Substrate using an ACF. *Microelectronics Reliability*, Vol. 45, No. 3-4, 2005, pp. 583-8. ISSN 0026-2714

Frisk, L. and Kokko, K. (2006). The Effects of Chip and Substrate Thickness on the Reliability of ACA Bonded Flip Chip Joints. *Soldering and Surface Mount Technology*, Vol. 18, No. 4, 2006, pp. 28-37. ISSN 0954-0911

Frisk L. and A. Seppälä A. (2006). Reliability of Flip Chip Joints on LCP and PI Substrate. *Soldering and Surface Mount Technology*, Vol. 18, No. 4, 2006, pp. 12-20. ISSN 0954-0911

Frisk L. and Kokko K. (2007). Effect of RCC on the reliability of adhesive flip chip joints. *Journal of Electronic Packaging*, Vol. 129 No. 3, 2007, pp. 260-5. ISSN 1043-7398

Frisk, L. and Cumini, A. "Effect of Substrate Material and Thickness on Reliability of ACA Bonded Flip Chip Joints", Soldering and Surface Mount Technology, Vol. 21, No. 3, 2009 pp. 15-23.

Frisk L. and Cumini A. (2010). Reliability of ACF interconnections on FR-4 substrates. *IEEE Transactions on Components and Packaging Technologies*, Vol. 33, No. 1, 2010, pp. 138-147. ISSN 1521-3331

Frisk, L., Saarinen K., and Kokko K. (2011). Reliability of ACA joined thinned chips on rigid substrates under humid conditions. *Proceedings of European Microelectronics and Packaging Conference*, EMPC, Brighton, UK, September 2011.

Ishibashi, K. and Kimura, J. (1996). A New Anisotropic Conductive Film with Arrayed Conductive Particles. *IEEE Transactions on Components, Packaging, and Manufacturing Technology*, Vol. 19, No. 4, 1996, pp.752–7. ISSN 2156-3950

Jang K-W., C-K. Chung, W-S. Lee, K-W.Paik. (2008). Material properties of anisotropic conductive films (ACFs) and their flip chip assembly reliability in NAND flash memory applications. *Microelectronics Reliability*, Vol. 48, No. 7, 2008, pp. 1052-61. ISSN 0026-2714

Jin, S., Tiefel, T.H, Li-Ilan C., and Dahringer, D.W. (1993). Anisotropically Conductive Polymer Films with a Uniform Dispersion of Particles. *IEEE Transactions on Components, Hybrids, and Manufacturing Technology*, Vol. 16, No. 8, 1993, pp.972 - 977. ISSN 0148-6411

Kim, J., Kwon, S., and Ihm, D. (2004). Reliability and Thermodynamic Studied of an Anisotropic Conductive Adhesive Film (ACAF) Prepared from Epoxy/Rubber Resins. *Journal of Material Processing Technology*, Vol. 152, No. 3, 2004, pp.357-62. ISSN 0924-0136

Kwon, W. and Paik, K. (2004). Contraction stress build-up of anisotropic conductive films (ACFs) for flip-chip interconnection: Effect of thermal and mechanical properties of ACFs. *Journal of Applied Polymer Science*, Vol. 93, No. 6, 2004, pp. 2634-41. ISSN 1097-4628

Kwon, W., Yim, M., Paik, K., Ham, S., and Lee, S. (2005). Thermal Cycling Reliability and Delamination of Anisotropic Conductive Adhesives Flip Chip on Organic Substrates with Emphasis on the Thermal Deformation. *Journal of Electronic Packaging*, Vol. 127, No. 2., 2005, pp.86-90. ISSN 1043-7398

Kwon, W., Ham, S., and Paik, K. (2006). Deformation Mechanism and its Effect on Electrical Conductivity of ACF Flip Chip Package Under Thermal Cycling Condition: An experimental Study. *Microelectronics Reliability*, Vol. 46, No.2-4, 2006, pp.589-99. ISSN 0026-2714

Kwon, W. and Paik, K. (2006). Experimental Analysis of Mechanical and Electrical Characteristics of Metal-coated Conductive Spheres for Anisotropic Conductive Adhesives. *IEEE Transactions on Components and Packaging Technologies*, Volume 29, No. 3, 2006, pp. 528-34. ISSN 1521-3331

Lai, Z. & Liu J. (1996). Anisotropically conductive adhesive flip-chip bonding on rigid and flexible printed circuit substrate. *IEEE Transactions on Components, Packaging, and Manufacturing Technology, Part B: Advanced Packaging*, Vol. 19, No. 3, 1996, pp. 644-66. ISSN 2156-3950

Lai, Z., Lai, R., Persson, K., and Liu, J. (1998). Effect of Bump Height on the Reliability of ACA Flip Chip Joining with FR4 Rigid and Polyimide Flexible Substrate. *Journal of Electronics Manufacturing*, Vol. 8, No. 3-4, 1998, pp. 217-24. ISSN 0960-3131

Lau, J. (1995). *Flip Chip Technologies*, McGraw-Hill, ISBN 0070366098, New York, USA.

Lau, J. (2000). *Low Cost Flip Chip Technologies for DCA, WLCSP, and PBGA Assemblies*, McGraw-Hill, ISBN 0071351418, New York.

Lau, J., Wong, C.P., Lee, N., and Lee, R. (2003). *Electronics Manufacturing: with Lead-Free, Halogen-Free, and Conductive-Adhesive Materials*. McGraw-Hill Professional, ISBN 0071386246, New York.

Li, Y. and Wong, C.P. (2006). Recent Advances of Conductive Adhesives as a Lead-free Alternative in Electronic Packaging: Material, Processing, Reliability and Applications. *Materials Science and Engineering R 51*, 2006, ISSN 0927-796X.

Licari, J. J. and Swanson, D. W. (2005). *Adhesives Technology for Electronic Applications*, William Andrew Publishing, ISBN 0815515138, Norwich, New York.

Liu, J., Tolvgard, A., Malmodin, J., and Lai, Z. (1999). A Reliable and Environmentally Friendly Packaging Technology - Flip Chip Joining Using Anisotropically Conductive Adhesive", *IEEE Transactions on Components and Packaging Technology*, Vol. 22, No. 2, 1999, pp. 186-90, ISSN 1521-3331.

Liu, J. (2001). ACA Bonding Technology for Low Cost Electronics Packaging Applications – Current Status and Remaining Challenges. *Soldering & Surface Mount Technology*, Vol. 13, No. 3, 2001, pp. 39-57, ISSN 0954-0911.

Lu, D. (2006). Overview of Recent Advances on Isotropic Conductive Adhesives. *Proceeding of 6th Conference on High Density Microsystem Design and Packaging and Component Failure Analysis (HDP)*, ISBN 9781424404889, Shanghai, China, June 2006.

Luo, S., Leisen, J., and Wong, C. (2002). Study on Mobility of Water and Polymer Chain in Epoxy and Its Influence on Adhesion. *Journal of Applied Polymer Science*, Vol. 85, No. 1, 2002, pp.1-8. ISSN 1097-4628

Mercado, Lei., white, J., Sarihan, V., and Lee, T. (2003). Failure Mechanism Study of Anisotropic Conductive Film (ACF) Packages. *IEEE Transactions on Components and Packaging Technologies*, Vol. 26, No. 3, 2003, pp. 509-16. ISSN 1521-3331

Morris, J., Lee, J., and, Liu, J. (2005). Isotropic Conductive Adhesive Interconnect Technology in Electronics Packaging Applications. *Proceedings of the 5th International Conference on Polymers and Adhesives in Microelectronics and Photonics*, ISBN 0780395530, Wrocaw, Poland, January 2006.

Ogunjimi, A. O., Mannan, S. H., Whalley, D. C., & Williams, D. J. (1996). Assembly of Planar Array Components Using Anisotropic Conducting Adhesives-A Benchmark Study: Part I-Experiment. *IEEE Transactions on Components, Packaging, and Manufacturing Technology*, Vol. 19, No 4, 1996, pp. 257-63. ISSN 2156-3950

Pinardi, K., Liu, J., Haug, R., Treutler, C., and Willander, M. (1998). Deformation study of the PCB during the flip chip assembly processing anisotropically conductive adhesive (ACA) as a bonding agent. Proceedings of 3rd IEEE Conference on Adhesive Joining and Coating Technology in Electronics Manufacturing, Binghamton, ISBN 0780349342, NY, USA, September, 1998.

Rizvi, M. J., Chan, Y. C., Bailey, C., Lu, H., and Sharif, A. (2005). The Effect of Curing on the Performance of ACF Bonded Chip-on-flex Assemblies After Thermal Ageing. *Soldering and Surface Mount Technology*, 2005, Vol. 17, No. 2, pp.40-8. ISSN 0954-0911

Saarinen, K., Frisk, L. and Ukkonen, L. (2011). Effects of different combinations of environmental test on the reliability of UHF RFID tags. *Procedings of 17th European Microelectronics and Packaging Conference, EMPC*, Brighton, UK September, 2011.

Savolainen, P. (2004). Display driver packaging: ACF reaching the limits?. *Proceedings of the 9th International Symposium on Advanced Packaging Materials: Processes, Properties and Interfaces*, ISBN 0780384369, Atlanta, Georgia, USA, January, 2004.

Suhir, E. (2002). Accelerated Life Testing (ALT) in Microelectronics and Photonics: Its role, Attributes, Challenges, Pitfalls, and Interaction with Qualification Tests. *Journal of Electronic Packaging*, Vol. 124, No. 3, 2002, pp. 281-91. ISSN 1043-7398

Sungwook, M. and Chappell, W. J. (2010). Novel Three-Dimensional Packaging Approaches Using Magnetically Aligned Anisotropic Conductive Adhesive for Microwave Applications. *IEEE Transactions on Microwave Theory and Techniques*, vol. 58, No. 12, pp. 3815-23. ISSN 0018-9480

Takagi, K., Honma, H., and Sasabe, T. (2003). Development of Sequential Build-up Multilayer Printed Wiring Boards in Japan. *IEEE Electrical Insulation Magazine*, Vol. 19, No. 5, 2003, pp. 27 - 56. ISSN 0883-7554

Tan, S. C., Chan, Y. C., Chiu, Y. W., and Tan, C. W. (2004). Thermal Stability Performance of Anisotropic Conductive Film at Different Bonding Temperatures. *Microelectronics Reliability*, Vol. 44, No. 3, 2004, pp. 495-503. ISSN 0026-2714

Uddin, M. A., Alam, M. O., Chan Y. C., and Chan, H. P. (2004). Adhesion Strength and Contact Resistance of Flip Chip on Flex Packages – Effect of Curing Degree of Anisotropic Conductive Film. *Microelectronics Reliability*, Vol. 44, No. 3, 2004, pp.505-14. ISSN 0026-2714

Wang, X. Wang, Y., Chen, G., Liu, J., and Lai, Z. (1998). Quantitative Estimate of the Characteristics of Conductive Particles in ACA by using Nano Indenter. *IEEE Transactions on Components and Packaging, and Manufacturing Technology*, Vol. 21, no. 2, 1998, pp.248 - 51. ISSN 1070-9886

Watanabe, I. Fujinawa, T. Arifuku, M. Fujii, M. Yasushi, G. (2004). Recent advances of interconnection technologies using anisotropic conductive films in flat panel display applications. *Proceedings of the 9th IEEE International symposium Advanced Packaging Materials: Processes, Properties and Interfaces*, ISBN 0780384369, Atlanta, Georgia, USA, August 2004.

Whalley, D., Mannan, S., and Williams, D. (1997). Anisotropic Conducting Adhesives for Electronic Assembly. *Assembly Automation*, Vol. 17, No. 1, 1997, pp.66-74. ISSN 0144-5154

Wong, E. H., Chan, K. C., Rajoo, R., and Lin, T. B. (2000). The Mechanics and Impact of Hygroscopic Swelling of Polymeric Materials in Electronic Packaging. *Proceedings of the 50th IEEE Electronics Components and Technology Conference*, ISBN 0780359089, Las Vegas, NV, USA, May, 2000.

Wu, S. X., Zhang, C. Yeh, C, Wille, S., and Wyatt, K. (1997). Cure Kinetics and Mechanical Properties of Conductive Adhesive. *Proceedings of the 47th IEEE Electronics Components and Technology Conference*, ISBN 0-7803-3857-X, San Jose, CA, USA, May, 1997.

Yim, M. J. and Paik, K. W. (1998). Design and understanding of anisotropic conductive films (ACF's) for LCD packaging. *IEEE Transactions on Components and Packaging Technologies*, Vol. 21, No. 2, 1998, pp.226-34. ISSN 1521-3331

Yim, M. J. and Paik, K. W. (1999). The Contact Resistance and Reliability of Anisotropically Conductive Film (ACF). *IEEE Transactions on Advanced Packaging*, Vol. 22, No. 2, 1999, pp.166-73, ISSN 1521-3323.

Yim, M. J. and Paik, K. W. (2001). Effect of Nonconducting Filler Additions on ACA Properties and the Reliability of ACA Flip Chip on Organic Substrates. *IEEE*

Transactions on Components and Packaging Technologies, Vol. 24, No. 1, 2001, pp.24-32. ISSN 1521-3331

Yin, C.Y., Alam, M.O., Chan, Y.C., Bailey, C. and Lu, H. (2003). The Effect of Reflow Process on the Contact Resistance and Reliability of Anisotropic Conductive Film Interconnection for Flip Chip on Flex Applications. *Journal of Microelectronics Reliability*, Vol. 43, No. 4, 2003, pp. 625-33. ISSN 0026-2714

Zhong, Z. W. (2005). Various Adhesives for Flip Chips. *Journal of Electronic Packaging*, Vol. 127, No. 1, 2005, pp. 29-32. ISSN 1043-7398

10

FE-SEM Characterization of Some Nanomaterial

A. Alyamani[1] and O. M. Lemine[2]
[1]National Nanotechnology Research Centre, KACST, Riyadh,
[2]Physics Department, College of Sciences, Imam University Riyadh,
Saudi Arabia

1. Introduction

In 1931 Max Knoll and Ernst Ruska at the university of Berlin built the first electron microscope that use accelerated electrons as a source instead of light source. However, the first scanning electron microscope (SEM) was built in 1938 due to the difficulties of scanning the electrons through the sample. Electron microscope is working exactly the same as the optical microscope expects it use a focused accelerated electron beam [1].

Since the invention of the electron microscope, it became one of the most useful instruments that has an impact in understanding scientific phenomena in different fields, such as physics, nanotechnology, medicine, chemistry biology..etc. Electron microscope has the ability to resolve objects ranging from part of nano-metre to micro-metre compared to light microscope that has a magnificationin the range of 1000 and resolution of 200 nm.

In the first part of the chapter, we will describe some of the basics of electronic microscope and its applications. The second part will be dedicated to the results obtained mainly by SEM.

2. Electron microscopy

2.1 Fundamental principles of electron microscopy

The principal of electron microscope is the same as a light microscope but instead of using visible light it use very energetic electrons as a source. However, the resolution of the optical microscope is limited by its wavelength compared to accelerated electrons which have very short wavelength.This is what makes it possible to see very small features.

In electron microscopes, electrons have very small wavelength λ. This wavelength can be changed according to the applied high voltage. Hence, according to Rayleigh's criterion the wavelength λ of an electron is related to the momentum $p=mv$ of the electron by: [2]

$$\lambda = \frac{h}{p} = \frac{h}{mv} \tag{1}$$

where $h = 6 \times 10^{-34}$ J s is the Planck constant, m and v are the mass and velocity of the electron respectively. Since the electron can reach nearly the velocity of light c then we can use the relativistic equations. In this case the electron mass is changing according to:

$$m = \frac{m_e}{\sqrt{1-(v/c)^2}} \tag{2}$$

where m_e is the rest mass of the electron. The energy eV transmitted to an electron is giving by:

$$eV = (m - m_2)c^2 \tag{3}$$

By using equations 1,2 and 3 the electron wavelength can be written as function of accelerated voltage: [3]

$$\lambda = \sqrt{\frac{1.5}{V\left(1+V*10^{-6}\right)}} nm \tag{4}$$

for example an accelrated voltage of 10 kV will yield a wavelength of 0.0122 nm. The extremely small wavelengths make it possible to see atomic structures using accelerated electrons.

2.2 Interaction of accelerated electrons with the specimen

The electron beam interacts with the specimen reveal useful information about the sample including: its surface features, size and shape of the features, composition and crystalline structure. The interaction of the electron beam with the specimen can be in different ways:

2.2.1 Secondary electrons

If the incident electronscome close enough to the atom then these electrons will give some of their energy to the specimen electrons mainly in the K-shell. As a result, these electrons will change their path and will ionize the electrons in the specimen atoms. These ionized electrons that escape the atoms are called secondary electrons. These electrons will move to the surface of the specimen and undergoing to elastics and inelastic collision until reaching the surface. However, due to their low energy $\sim 5eV$ only those electrons that are close to the surface ($\sim 10\ nm$) will escape the surface and then can be detected and can used for imaging the topography of the specimen.

2.2.2 Backscattered electrons

When the incident electrons hit an atom directly, then they will be reflected or back-scattered. Different atomic type of atoms will result in a different rate of backscattered electrons and hence the contrast of the image will vary as the atomic number of the specimen change, usually atoms with higher atomic number will appear brighter than those have lower atomic number.

2.2.3 Transmitted electrons

If the incident electrons pass through the specimen without any interaction with their atoms, then these electrons called transmitted electrons, these electrons are used to get an image of thin specimen. Another scattering mechanism called elastic scattered where electrons don't loss their energy these scattered electrons can be used to get information about orientation and arrangement of atoms.

2.2.4 Other interactions

When the atoms bombarded with incident electrons, electrons will released from these atoms and this will leave the atom in the excited state. In order for the atom to return to the ground state, it needs to release the excess energy Auger electrons, X-Rays,and cathodoluminescence are three ways of relaxation. The x-ray is used to identify the elements and their concentrations in the specimen by using a technique called Energy –dispersive X-ray analysis (EDX) technique. Chemical analysis can be done by using Auger electrons.

2.3 Types of electron microscopes

There are two types of electron microscopes. Scanning Electron Microscopes (SEM), and Transmission Electron Microscope (TEM), these types of microscopes detect electrons that emitted from the surface of the sample. The accelerated voltage is ranging from 10kV to 40kV for the SEM. The thickness of the specimen in this case is not important. In addition, the samples to be tested have to be electrically conductive; otherwise they would be overcharged with electrons. However, they can be coated with a conductive layer of metal or carbon.

In TEM the transmitted electrons are detected, and in this case the specimen thickness is important and typically should not exceed 150 nm. The accelerated voltage in this case $\geq$ 100kV.

Since the electrons are easily scattered in air all electron microscopes should operate under a high vacuum.

All types of electron microscopes are basically consist of three basic components:

Electron Gun which is used to provide and supply electrons with the required energy.There are different types of electron gun;the old type was a bent piece of Tungsten wire with 100 micro-metresin diameter. Higher performance electron emitters consist of either single crystals of lanthanum hexaboride (LaB6) or from field emission guns.

3. Experimental

3.1 Pulse Laser Deposition (PLD)

As a materials processing technique, laser ablation was utilized for the first time in the 1960's, after the first commercial ruby laser was invented [4]. Nevertheless, as a thin film growth method it did not attract much research interest until the late 1980´s [5], when it has been used for growing high temperature superconductor films. Since then, the development of the pulsed laser deposition (PLD) technique has been more rapid and the amount of research devoted to this topic has increased dramatically [6]. The growth and quality of the resulting film will generally depend on a number of fundamental parameters, including the choice of substrate, the substrate temperature and the absolute and relative kinetic energies and/or arrival rates of the various constituents within the plume.

The PLD process is shown in figure 1:

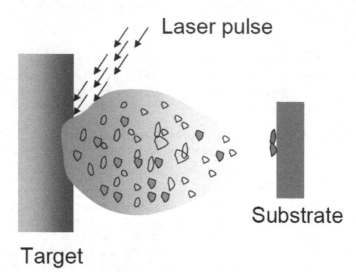

Fig. 1. Schematic presentation of the pulsed laser deposition process
a) Laser − target interaction, b) Plume expansion and c) Film deposition [6].

The growth and quality of the resulting thin film will generally depend on a number of fundamental parameters, including the choice of substrate, the substrate temperature, T_S, distance between target-substrate, pressure and laser energy.

In our case the laser energy was 300 mJ and the time was fixed at 60 minutes. For the others parameters (substrate, substrate temperature, pressure), different values were used.

3.2 Mechanical Alloying (MA)

The ball milling constitutes new promising methods to produce nanosized particles [7,8]. It has many advantages, e.g low cost, simple operation.The ball-milling is generally used as a mechanical co-grinding of powders, Initially different in nature, up to the preparation of a new powder, homogeneous in composition. The milling is done in cylindrical containers called vials and containing balls. The nature of the milling tools can be as diverse as steel, agate, tungsten carbide... The vials are generally filled under an inert   atmosphere to avoid side reactions, since the particles are fractured during the milling process and, therefore, new highly reactive surfaces can react with the surrounding gases [8].

Several terms are used to call this technique: "Mechanical Alloying" when there is a chemical reaction between different powders, "Mechanical Grinding" or "Mechanical Milling" when the only goal is to modify the texture and/or the structure of a material (no chemical reaction is involved in the process).

Two kinds of milling systems were used to prepare our nanopowders (Vibrant and planetary milling) and different milling parameters were considered (milling times, balls to powders mass ratio, size of balls and rotation speed).

3.3 Filed Emission Scanning Electron Microscopy (FESEM)

The field emission scanning electron microscope (FE-SEM) is a type of electron microscope that images the sample surface by scanning it with a high-energy beam of electrons in a raster scan pattern. Electron emitters from field emission gun was used. These types of electron emitters can produce up to 1000x the emission of a tungsten filament. However, they required much higher vacuum conditions. After the electrons beam exit the electron gun, they then confined and focused into a thin focused, monochromatic beam using metal aperturesand magnetic lenses. Finally, Detectors of each type of electrons are placed in the microscopes that collect signals to produce an image of the specimen.

Particles morphology of our samples was investigated using Nova 200 NanoLab field emission scanning electron microscope (FE-SEM).

4. Results

4.1 Thin film prepared by Pulse Laser Deposition (PLD)

Fig. 2 shows FESEM micrographs of ZnO thin films grown on sapphire substrate by pulse laser deposition at growth temperature from 685 to 750 °C by using a ZnO powder target at high grade. The experimental parameters are summarized in table.1. It is seen that with the substrate temperature increasing the morphology of ZnO thin films have a little difference. The thickness of films decreases with the increase of substrate temperature.

The effect of the distance between target and substrate on the morphology was also studied. Fig. 3 shows the FESEM images of ZnO thin film with different distance between the target and thin film. It is clear that the distance affect the morphology of the film.

TEMPERATURE(°C)	THICKNESS (nm)	Distance between target and the film	Oxygen pressure	LASER ENERGY(mJ)	SUBSTRATE
750	510	37.5 mm	150 mTorr	350	Sapphire
700	1230	37.5 mm	150 mTorr	350	Sapphire
685	1115	37.5 mm	150 mTorr	350	Sapphire
400	-	10 mm	150 mTorr	350	Sapphire
400	-	23mm	150 mTorr	350	Sapphire

Table.1.Growth parameters of ZnO thin films

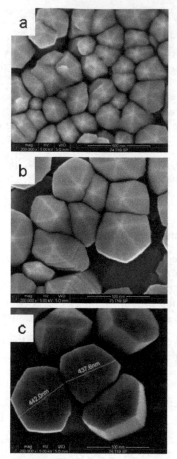

Fig. 2. FESEM images of thin film grown on sapphire at substrate temperature of: (a) 750 °C, (b) 700 °C and (c) 685°C.

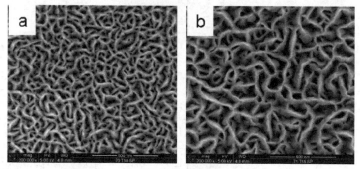

Fig. 3. **FESEM** images of thin film grown on sapphire at substrate temperature of 400°C: a) distance between target and thin film =10mm and b) distance between target and thin film =23 mm

4.2 Nanopwders obtained by mechanical alloying

4.2.1 Hematite (α-Fe₂O₃) nanocrystallines

The conditions for production of α-Fe$_2$O$_3$ nano-crystallines by dry milling was studied. [9,10] Commercial α-Fe$_2$O$_3$ powder was used as the starting material. The mechanical milling was carried out in a planetary ball mill Fritsch Pulverisstte 6. The powder was ground in vial with 200g of mixture 1:1 in weight of stainless steel balls (10 and 15 mm in diameter). Different milling times were considered (1, 6, 12, 24 and 48h) and the sample to balls weight ratio was fixed to 1:10. The milling intensity was 250 rpm. Fig.4 shows scanning electron micrographs before and after milling. It is clear that un-milled powder shows an in homogeneities regarding particle size distribution (Fig. 4a). After milling, a reduction of the particle size can be observed with relatively better homogeneity (Fig. 4b-d)). SEM images for increasing milling times reveal clearly that large particles are in fact agglomerates of much smaller particles.

Fig. 4. FESEM images for the samples milled at different times : a) 0h, b) 12h, c) 24h and d) 48h.

4.2.2 Nanocrystalline zinc ferrite (ZnFe₂O₄)

Nanocrystalline zinc ferrite (ZnFe$_2$O$_4$) is synthesized by high-energy ball-milling from a powders mixture of zinc oxide (ZnO) and hematite (α-Fe$_2$O$_3$). [11] Commercially powders of hematite (α-Fe$_2$O$_3$) and zinc oxide (ZnO) are used with equal molar (1:1) and were introduced into a stainless steel vials with stainless steel balls (12 mm and 6 mm in diameter) in a high energy mill (SPEX 8000 mixer mill). Different milling times were considered (6, 12 and 24) and two values of the balls to powders mass ratio were used (10:1 and 20:1). SEM micrographs of the samples before and after milling are shown in Figure 5 It is clear that unmilled powder shows a different chap of powders due to zinc oxides and hematite powders (fig.5a, 5b). After milling, a reduction of the crystallite size can be observed fig.5c. High magnification images (fig.5d) reveal clearly the formation of a new nanocrystalline different from the started materials.

4.2.3 Zinc oxides Nanocrystalline (ZnO)

The effects of milling times on the mechanically milled ZnO powder are also studied [12]. Commercially ZnO powders with average particle size of about 1 μm and 99.9% of purity, were introduced into a stainless steel vials with stainless steel balls (12mm and 6mm in diameter) in a SPEX 8000 mixer mill, then milled for different milling periods of time. The balls to powder mass ratio was fixed to 10:1. SEM micrographs of the samples before after milling are shown in Figure.6. It is clear that un-milled powder shows un-homogeneities regarding particle size distribution, where the average size varies in the range 150 – 800 nm (Fig. 6a, 6b). After milling, a reduction of the particle size can be observed with relatively better homogeneity (Fig. 6c, 6e). High magnification images (Fig. 6d, 6f) reveal clearly that large particles are in fact agglomerates of much smaller particles. The average particle size after milling is less than 100nm.

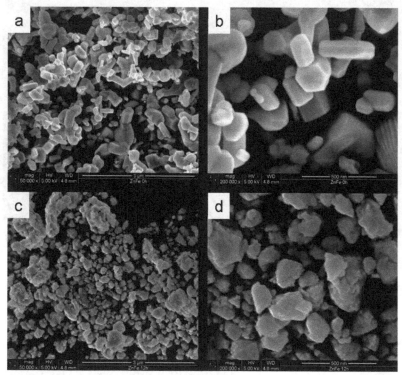

Fig. 5. FESEM micrographs of mixtures (zinc oxides +hematite) powders as received (a) as received at high magnification (b) milled for12h (c) milled for 12h high magnification (d)

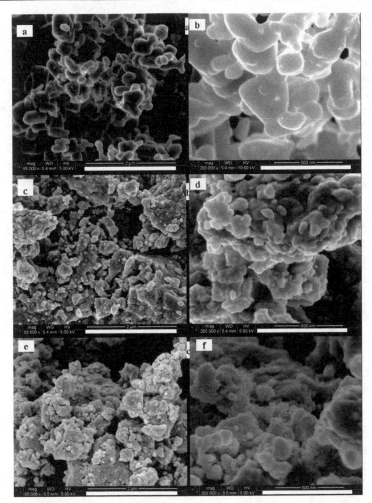

Fig. 6. SEM micrographs of ZnO powder for different milling time: a) as received; b) as received at high magnification; c) milled for 3h; d) milled for 3h high magnification; e) milled for 5h; f) milled for 5h at high magnification.

In summary, it is clear that scanning electron microscopy gives tremendous information about the microstructure of nanomaterials including thin film and nano-powders. In addition to that the signals coming from the sample can be used to get information about the composition of the materials and the structure.

5. References

[1] David B Williams and C Barry Carter, "Transmission Electron Microscopy", Springer 2009
[2] Nouredine Zietteli, " Quantum mechanic, concept and applications", Wiley 2001.

[3] Arthur Beiser, " Concept of modern physics", McGraw-Hill, Inc, 1995 H.M. Smith and
    A.F. Turner, Appl. Opt. 4 (1965) 147.
[4] D. Dijkkamp, T. Venkatesan, X.D. Wu, S.A. Shaheen, N. Jisrawi, Y.H. Min-Lee, W.L.
    McLean and M. Croft, Appl. Phys. Lett. 51 (1987) 619-621.
[5] D.B. Chrisey and G.K. Hubler (Eds.), "Pulsed Laser Deposition of Thin Films", Wiley,
    New York, 1994.
[6] Raphaël Janot and Daniel Guérard, Progress in Materials Science, Volume 50, Issue 1,
    January 2005, 1-92
[7] E. Petrovsky, M.D. Alcala, J.M. Criado, T. Grygar, A. Kapicka and J. Subrt, J. *Magn.
    Magn. Mater.* 210 (2000), p. 257.
[8] O.M.Lemine., A.Alyamani, M. Sajieddine and M.Bououdina,, Journal of alloys and
    compounds, 502 (2010), pp. 279-282
[9] O. M. Lemine , R. Msalam, M. Sajieddine , S. Mufti, A. Alyemani , A. F. Salem, Kh. Ziq
    and M. Bououdina, International Journal of Nanoscience, Vol. 8, No. 3 (2009) 1–8.
[10] O.M. Lemine , M. Bououdina, M. Sajieddine, A. M. Al-Saie, M. Shafi, A. Khatab, M. Al-
    hilali1 and M. Henini, Physica B 406 (2011) 1989–1994
[11] O. M. Lemine, A.Alyemani and M.Bououdina, Int. J. Nanoparticles, Vol. 2, 2009

Morphological and Photovoltaic Studies of TiO$_2$ NTs for High Efficiency Solar Cells

Mukul Dubey and Hongshan He[*]

Center for Advanced Photovoltaics, Department of Electrical Engineering & Computer Science South Dakota State University, Brookings, SD, USA

1. Introduction

Highly ordered nanostructures, especially TiO$_2$ NTs, have attracted considerable research interest in recent years due to their diverse applications in photocatalysis, photonic crystals, sensors, batteries and photovoltaic devices. The photophysical, photochemical, electrical and surface properties of these nanostructured materials depend highly on their morphology because of the quantum size effect. Hence it is critical to study the effect of morphology of the ordered nanostructures for device applications. In this chapter we will only focus on the TiO$_2$ NT morphology in context of their applications in dye-sensitized solar cells (DSCs).

DSC is an electrochemical device that converts sunlight to electricity. The major components of DSC are photoelectrode, counterelectrode and electrolyte sandwiched between them. The photoelectrode is a dye-coated wide band gap semiconductor, such as TiO$_2$, on a transparent conducting oxide (TCO) glass substrate. Dye molecules absorb sunlight and the electrons in the ground state are excited to the excited state. The electrons in the excited states inject into the conduction band of TiO$_2$. The injected electrons transports to the TCO electrode via diffusion through TiO$_2$ NPs. The electrons then flow through the external circuit to the counterelectrode, which is usually a platinized TCO glass. The redox species in the electrolyte, usually iodide, take the electron from counterelectrode, and are reduced to tri-iodide, which further gets oxidized by providing its electron to the ground state of dye molecule for its regeneration. There are several factors that affect the efficiency of DSC such as absorption band of dye molecule, electron injection efficiency from dye to TiO$_2$, redox potential of electrolyte and charge transport through TiO$_2$. The morphology of TiO$_2$ photoelectrode is one critical factor that plays a pivotal role in the conversion of sunlight to electricity in DSCs. Remarkable breakthrough in photoelectrode by changing the planar structure to randomly packed mesoporous structure of TiO$_2$ NPs improved the efficiency from less than 1 % to 8% by Grätzel *et al*. The mesoporous structures are promising due to their high surface area for the adsorption of photosensitzer leading to the improved light absorption and hence high efficiency. The photoelectrode was further optimized by introducing a compact layer with small TiO$_2$ NPs and a scattering layer with large TiO$_2$ NP underneath and at the top

[*] Corresponding Author

of normal TiO$_2$ NPs respectively. Both improved electrical and optical properties of photoelectrode and hence the device efficiency. With those structures and ruthenium bipyridine dyes, a respectable efficiency of 11.5% has been achieved rendering the DSCs as promising and cost-effective alternative to its otherwise expensive silicon technology.

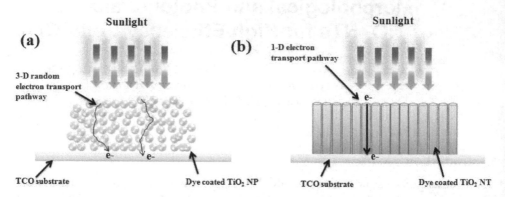

Fig. 1. (a) Schematic representation of electron transport in TiO$_2$ NPs based photoelctrode; (b) electron transport in TiO$_2$ NT based photoelectrode

The electron collection efficiency is a critical factor governing the overall photo conversion efficiency of solar cell. Various investigations suggest that the random morphology of polycrystalline TiO$_2$ NPs exhibits high defect density, which leads to the electron losses via recombination and the reduced electron collection efficiency. The presence of numerous defects, grain boundaries and surface states provides several trapping/detrapping and recombination sites in the electron transport pathway. The presence of defects reduces the electron mobility leading to increased recombination and hence reduced cell performance. In this regard anodic TiO$_2$ NTs proposed by Grimes *et al* is considered as an excellent electron acceptor for DSC. Architecturally, these NTs are well aligned in regular array perpendicular to the substrate leading to rapid unidirectional electron transport with reduced recombination. A schematic for difference in dimensionality of electron transport between random nanocrystalline particle network and one-dimensional NT is shown in Figure 1. The electron from dye molecules migrate directly from top of the NT to the bottom for electron collection without migration in a three dimensional network. A close to 100% electron collection efficiency at the bottom of the nantotube was observed. In addition, NTs also have strong light scattering behavior which increases the optical path length in the film and improve the light absorption efficiency for high solar cell efficiency.

Despite being promising both electrically and optically, the highest energy conversion efficiency obtained from NT based DSCs is only ~ 7%, which is much lower than the conventional NP based DSC. One of the disadvantages identified was the back illumination geometry of devices due to the presence of non-transparent Ti metal underneath the TiO$_2$ NT arrays. The TiO$_2$ NT arrays are usually grown directly from a thin layer of Ti metal, which is difficult to remove. This requires photo illumination from the counterelectrode (a platinum coated transparent conducting electrode) side as shown in Figure 2. The back illumination leads to significant loss in the photon flux by reflection from the platinum and absorption in the electrolyte. It was difficult to realize front illumination since the NTs were grown on titanium substrate and no technique was known to either grow or transfer the NT films on to the transparent conducting substrate.

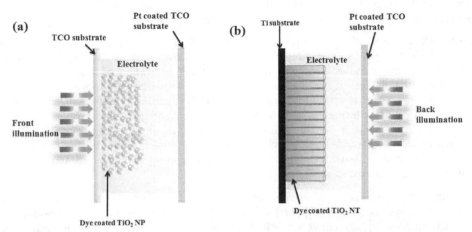

Fig. 2. **(a)** TiO$_2$ NP based DSC with front illumination geometry from photo-electrode side; **(b)** TiO$_2$ NT based DSC with back illumination geometry from counter-electrode side

Front illumination in TiO$_2$ NT-based DSCs can be realized through several recently reported methods. The first method is the growth of NTs on glass substrate with sputtered Ti metal on top. The sputtering must be performed at high temperatures to prevent peeling after anodization. Grimes *et al* recently reported a new method for sputtering Ti on FTO glass at low temperature that produced TiO$_2$ NTs with lengths up to 33 μm after anodization. A cell with a 17 μm NT array achieved a conversion efficiency of 6.9%. Two concerns emerge with this process: (1) the time- consuming nature of sputtering several tens of micrometer Ti may increase cost, and (2) the FTO layer on the glass could be damaged during anodization.

The second method is to remove the NT array from the Ti foil and attach it to FTO glass. In 2008, Jong Hyeok Park *et al* put anodized Ti foil in 0.1 M HCl aqueous solution for 1 hour, obtained an NT membrane, and attached it to FTO glass with the help of titanium isopropoxide. They achieved 7.6% efficiency with 8 μm NT arrays. Although the team claimed that NT membranes could be handled with tweezers, optical images in their publication suggested that these NT membranes were very fragile. In 2009, Qinwei Chen *et al* reported a re-anodization process that was followed by immersing the foil in 10% aqueous H$_2$O$_2$ solution for 24 hours and resulted in large sized NT membranes. The NTs were then attached to FTO glass with the help of a TiO$_2$ NP paste to achieve a conversion efficiency of 5.5%. Long-time immersion in solution diminishes the attractiveness of this mild process.

He *et al* also developed a method that can lift off the NT arrays in less than four minutes. The yellow membrane could easily be transferred to other substrates without any fracturing. He *et al* also developed a unique low temperature method to tightly plant the NT membranes on FTO glass. The NTs were embedded inside the NP layer. The DSCs with these films exhibited 6.1% efficiency using N719 as dye. It was found that the geometry of NT orientation on the glass substrate also plays a significant role in determining the efficiency of DSC. The test tube geometry of NTs with one end open and other end closed provides freedom in choosing the configuration of the freestanding NT fixation on substrate with either closed end or the open end on to the substrate. This finding suggests that both optically and electrically open end of the NT on to the substrate is superior to the other orientation and hence can help significantly in improving DSC efficiency.

Another challenge for the effective use of NT for DSC application is how to grow highly ordered TiO_2 NT arrays. Many researchers have reported that the NT tends to cluster together and form bundles which not only inhibits the infiltration of dye and electrolyte throughout the thickness of film but also increases recombination by incorporating disorder induced defects. It was reported that fine polishing of the titanium substrate prior to growth minimized the cluster formation. Several reports also indicated that the bundle and micro crack formation in the film was due to the capillary stress during the sample drying process. The supercritical CO_2 oxide drying technique was introduced, which indeed reduced the formation of clusters; however, the complete understanding of cluster formation is still elusive and requires further study.

To summarize the morphology of TiO_2 NT plays a critical role in dye-sensitized solar cell. Study of the effect of morphology of TiO_2 NT on DSC performance is therefore worthy of pursuit for achieving high conversion efficiency of the DSCs. In the following sections, we will discuss the growth mechanism of TiO_2 NTs and approaches for highly ordered TiO_2 NT array of NTs for DSC applications. We will also discuss how the effect of orientation of the NT on the TCO glass affects the photovoltaic properties of DSC.

2. Growth mechanism of TiO_2 NTs

This section reviews the growth mechanism of TiO_2 NTs by potentiostatic anodization technique in fluoride-containing electrolyte. The NT formation in acidic electrolyte containing F- ion is generally agreed to occur via the field assisted formation and dissolution of oxidized titanium surface. It involves two critical steps that occur simultaneously: formation of TiO_2 on the titanium surface and the dissolution of oxide. The process can be described by following two reactions:

$$Ti + 2H_2O \rightarrow TiO_2 + 4e^- + 4H^+ \dots\dots\dots\dots\dots(1)(Oxidation)$$

$$TiO_2 + 6F^- + 4H^+ \rightarrow [TiF_6]^{2-} + 2H_2O \dots\dots\dots(2)(Dissolution)$$

In this two-electrode setup, titanium serves as anode and platinum as cathode. The electrolyte is composed of ethylene glycol, ammonium fluoride and water. A constant DC voltage is applied across the electrodes as shown in Figure 3. After some time, a layer of TiO_2 NTs will form on the surface of Ti metal. Figure 4 shows schematically how the TiO_2 NTs are formed. When pristine Ti is immersed into electrolyte solution, it is surrounded by various ionic species such as OH- and F-. (a) Once the DC voltage is applied these ionic species tends to oxidize the surface of titanium substrate (b) forming a thin barrier layer of TiO_2 as depicted in the equation 1 of reaction mechanism. Simultaneously the process of dissolution of TiO_2 layer in presence of F- ion occurs leading to the formation of random pores during the initial stage of growth process (c). The F- ions localize to the bottom of the pore i.e. at the oxide/metal interface which further undergoes oxidation and dissolution processes. Since the concentration of F- ion is more at the bottom of the pore due to the external electric field; the effective dissolution of TiO_2 is more pronounced at the pore bottom leading to vertical cavity formation (d to f). The formation of round shape at the bottom of tube is still a topic of debate. It is proposed that this is results of volume expansion of TiO₂ compared to the space available from metal loss leading to high stress at the interface, high electric field distribution density at pore bottom and enhancement in acidity at the pore bottom due to the external electric field.

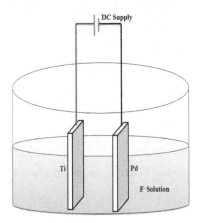

Fig. 3. Electrochemical anodization set up.

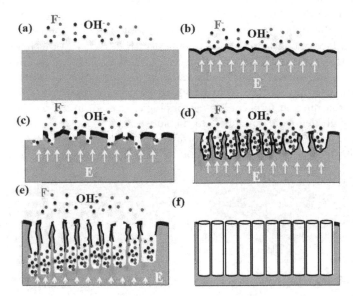

Fig. 4. (a) Titanium substrate in the ionic environment of electrolyte; (b) Formation of porous oxide layer on exposed surface of titanium right after field is switched on; (c) Initial random pore growth by dissolution; (d) elongation of pore geometry after few minutes of anodization; (e) development of regular array of pore geometry in the field direction; (f) fully developed NT array. Red and Black dots represent the fluoride & hydroxide ions respectively.

3. Effect of substrate morphology on growth of TiO₂ NTs

The formation of NTs largely depends on the type and concentration of ionic species present in the electrolyte as well as the extrinsic parameters such as anodization voltage, time and temperature. By controlling these factors, TiO_2 NTs having different length, diameter, and wall thickness can be obtained. However, it should be noted that field assisted directional dissolution of the oxide layer formed on titanium foil is a crucial step towards the formation of NTs which so far have been shown to depend on many variables such as electrolyte composition, concentration, anodization voltage and time, but least importance was given to the effect of substrate morphology on the growth of NTs which is discussed in the next section. We found that the morphology of titanium substrate also plays a key role in the morphological order of the NT thus formed. This section highlights the effect of morphological features of titanium substrate on NT growth which is further connected with the microscopic morphology drawing outline for the plausible reasons for the clustering of NTs and cost effective way to deal with it.

3.1 Effect of mechanical treatment of titanium substrate on TiO₂ NT growth

Commercial Ti foil with thickness ~ 250 μm is usually used for the growth of the TiO_2 NTs arrays. Before the anodization the Ti foil is cleaned by detergent, ethanol, toluene, and deionized water sequentially to remove any impurities on the surface. There are several commercial providers for Ti foil with high purity; however, the surface morphology of these as-purchased Ti foils is quite different. It was found the as-purchased Ti foil has many crack sites distributed throughout the surface of the substrate. Figure 5 shows the typical SEM image of the surface of one sample from Sigma-Aldrich. Many cracks were observed on the surface. The size of the cracks ranges from several hundred nanometer to several micrometer. The presence of such cracks leads to the formation of vertical gaps on the substrate leading to the absence of material up till certain depth. In addition there are several submicron range heterogeneous morphologies present in the vicinity of crack sites which render high degree of roughness to the substrate. The existence of cracks on the Ti surface leads to high degree of non-uniformity in the morphology of NTs thus formed resulting in the cluster and bundle formation of NT. Figure 6 (a) shows the SEM image of

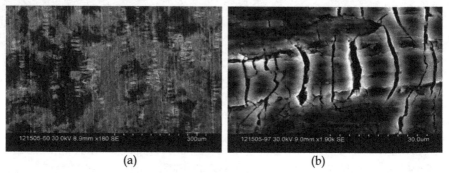

(a)   (b)

Fig. 5. **(a)** Cracks or vertical gaps present on the surface of as purchased commercially pure titanium substrate; **(b)** magnified image of crack showing the absence of material up till certain depth.

surface morphology for NTs grown on as purchased commercially pure titanium foil for 15 minutes. The fingerprint of substrate crack structures and submicron heterogeneously distributed morphology near crack site were clearly observed on the NT film.

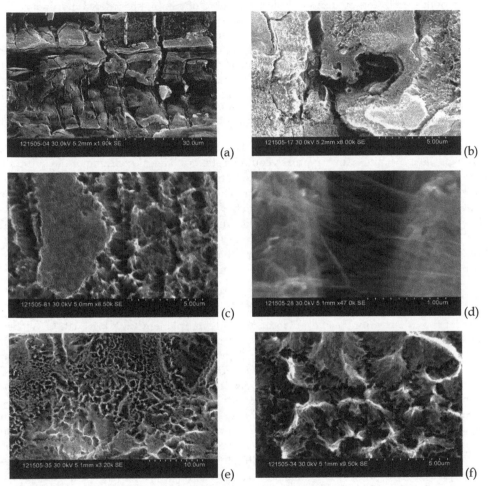

Fig. 6. (a) TiO₂ NTs after 30 min anodization; (b) TiO₂ NT with a whirlpool geometry at crack site; (c) TiO₂ NT cluster formed at crack lines; (d) Collapsed TiO₂ NTs at crack lines; (e) TiO₂ bundles throughout sample and (f) TiO₂ bundles under higher magnification

We further investigated the local morphology of NTs near the crack sites which is shown in higher magnification SEM image of Figure 6 (b). Whirlpool geometry of NT distribution at the crack site was observed, which shows the strong influence of substrate morphology on the initial growth of NTs. This effect was more pronounced in the NT under short anodization time. The clusters are formed near the crack lines of the substrate. Uniformed NTs are observed on the surface without any cracks. We also observed that the tubes over the edges of cracks tended to collapse on each other forming intercrossed tubes as shown in

Figure 6 (d). The collapsing of the NTs on each other can potentially lead to the cluster formation which can be seen from Figure 6 (e & f).

Based on the results of anodization on commercially purchased Titanium, it can be observed that smooth surface for anodization is very crucial to obtain highly ordered morphology of NTs. Han et al and Lee et al reported two step anodization processes to obtain ordered morphology of NTs. In their report first anodization was performed for shorter time followed by removal of the first NT layer. The surface of Ti after removal of first layer was very smooth leading to highly ordered morphology of NT formed in the second step. On the other hand Kang et al reported electropolishing technique in which Ti substrate was electropolished to render it a smooth surface followed by anodization to form ordered NT structure. Both electropolishing and two step anodization processes were found promising to obtained highly ordered NT array.

However, these processes involves complex two step processes which is time consuming and expensive. An alternative approach could be the mechanical polishing of the substrate to remove cracks. To this end we have tried to polish the Ti substrate using fine sand paper. However, our SEM results shows that even with very fine sand paper the micron size scratches are developed on the surface scratches are developed on the surface. It can be clearly seen that there It can be clearly seen that there were significant clumping and clustering of the NTs. Additionally at many other places NTs were found to be completely broken. Based on the results it can be inferred that even the fine mechanical polishing can form micron level roughness which cannot be used to grow highly ordered NTs.

3.2 Effect of chemical treatment of titanium substrate on TiO_2 NT morphology

In order to further verify the effect of local substrate morphology on NT growth, we etched the titanium substrate for 30 minutes in 0.75 M hydro fluoric acid (HF) introducing high degree of surface roughness to the substrate. Figure 8 (a) shows the morphology of rough surface of titanium after etching. TiO_2 NTs were then grown on the etched substrate for 15 minutes. It was observed that the initial pore formation for NT growth takes the local geometry of the substrate as shown in Figure 8 (b). The local pore formation might largely depend on the direction of local electric field was further confirmed by the NT formation in the etched substrate. Figure 8 (c) shows the SEM image of a large pit formed on the substrate due to etching. The pit shown in the image can be visualized to have three different planes i.e. x-y, y-z and x-z. It is interesting to note that the pore formation can be seen on all these three planes with their cross-sections perpendicular to the respective plane clearly indicating that the initial pore formation does depend on the direction of local electric field at the breakdown site this further depends on the local morphology of the substrate as shown in Figure 8 (d). The dependence of NT growth associated with the local electric field distribution corresponding to the substrate morphology can be a profound reason for the bundle and cluster formation in NTs which was further confirmed from SEM results. Figure 8 (e) shows the SEM image of NT at one of the crack sites of the NT film grown on etched substrate. It can be clearly observed that the NTs at crack site grew in different direction. Considering x-y plane to be the plane of substrate and z as direction normal to the substrate which is the preferred direction of NT growth, it can be clearly seen that the cross-sectional plane of NTs are facing in two different directions, one parallel to x-y plane highlighted with red circle and other in z- direction highlighted with yellow circle. The NTs facing x-y direction bends toward the z- direction. The initial bending followed by z growth of NTs was further confirmed in Figure 6 (f) where it can be observed from one of the pits that the initial pore formation on the walls of the pit is in all three directions. However as the NTs grew longer they start bending in one direction which latter completely follows one directional growth. Interestingly it can be seen that the initial bending ranging to several microns leads to the collapse of NTs on each other leading to the formation of clusters. Hence formation of highly ordered NTs can be severely influenced by the substrate morphology.

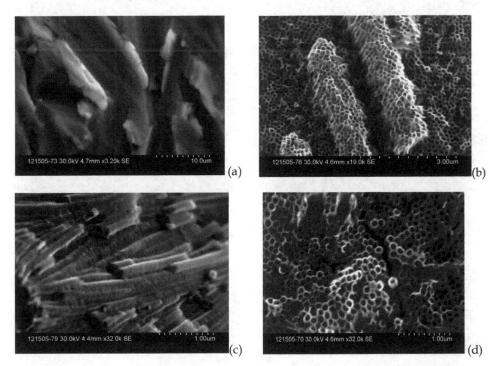

Fig. 7. (a) SEM image of polished Ti substrate; (b) bundle formation and non-uniformed TiO2 NT morphology; (c) side view showing different length of NTs and the bundle formation; (d) unevenly packed TiO2 NTs

Removing structural disorder from NTs was recently a key concern in the area of DSC. Some techniques including post growth ultrasonic treatment and supercritical CO_2 drying of NT samples showed promise in removing of the structural disorder. These techniques are very useful if the disorder in NT morphology is induced through impurities in the electrolyte, viscosity of the electrolyte or during drying of NTs after growth. Their applications to remove substrate induced disorder are limited. We employed a chemical etching process to solve this problem. The Ti substrates were immersed in 0.75 M HF ranging from 1 to 15 minutes. The cracks present on the substrate were removed completely in 10 minutes of etching time. Figure 9 (a) shows the SEM image of titanium foil etched for 5 minutes in 0.75 M HF where the crack features could still be observed. Figure 9 (b) shows

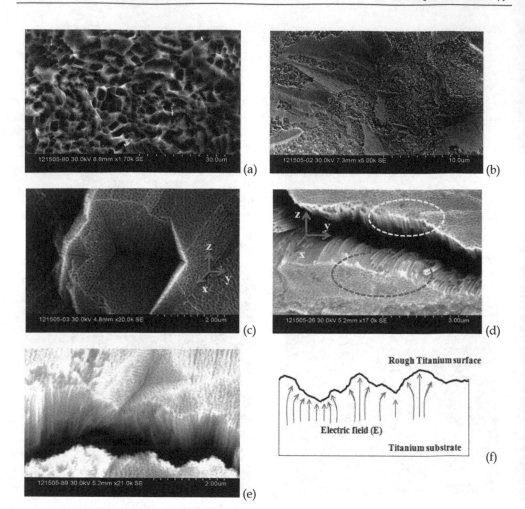

Fig. 8. SEM images of (a) etched Ti substrate; (b) TiO2 NTs grown on etched Ti; (c) large pit of TiO2 NTs on etched Ti; (d) TiO2 NT at a crack site on etched Ti; (e) TiO2 NT at the edge of one pit (f) non uniform local electric field distribution near the rough surface of titanium.

the cracks or vertical gaps completely disappeared after 10 minutes of etching but also introducing high degree of surface roughness induced on the substrate. Further etching the substrate for 15 minutes led to highly disordered coarse surface as can be seen in Figure 9 (c). A closer investigation of individual pits formed after 10 minutes etching of the substrate as shown in Figure 9 (d) revealed that these pits offer a very smooth concave shaped surface with average size of 5 – 10 μm. This observation suggested that highly oriented NTs can be grown over these smooth surfaces with short range of order on the surface of the substrate. Further concavity of the pit structure can lead to small bending in the NTs with cross-section plane facing towards the center of conic cross-section. The small bending of NTs can further help preventing the NTs to interact and collapse over the NTs formed in the neighboring pits, providing global order in the overall morphology of NTs. In order to verify our assumption we performed 30 minutes of anodization to grow shorter NTs on the titanium substrate etched for 10 minutes in 0.75 M HF. Figure 10 (a) shows the SEM image of NTs grown on etched substrate for 30 minutes anodization time. The image clearly shows that the NTs followed the local morphology of each pit taking the overall geometry of the substrate. In addition clustering or collapse of NTs was also not observed anywhere on the surface suggesting that overall order in the morphology can be achieved by this process. However, the method can find its applicability only when longer NTs can be successfully grown with long range order which is the essential need for solar cells. To investigate the morphology of longer NTs, we performed anodization of the etched substrate for 5 hrs which can lead to the formation of ~ 20 μm long NTs.

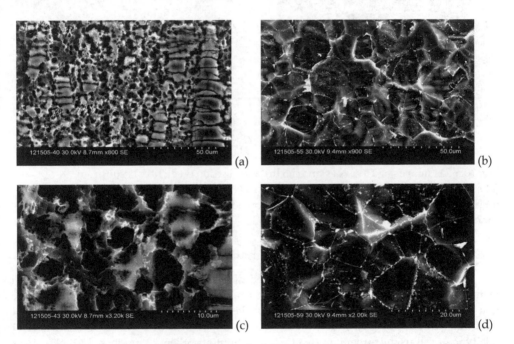

Fig. 9. SEM images of titanium substrate etched in 0.75 M HF under different etching time. (a) 5 minutes; (b, d) 10 minutes; (c) 15 minutes.

Interestingly the SEM image of Figure 10 (b) shows that the NTs even after 5hrs of anodization time followed highly ordered morphology without cluster formation anywhere on the substrate. It was also evidenced that the NTs retained the concave geometry of the substrate shown highlighted in yellow circle of Figure 10 (c). The overall morphology of the NTs were observed to be comprised of several small concave shaped honeycomb structure grouped together to form structured NT film which can be seen from SEM image of Figure 10 (d). Thus it can be seen that the morphology of the NTs significantly depends on both the morphology of the substrate and simple chemical pretreatment of the substrate can prove to be useful in growing oriented NTs which might further help in improving the efficiency of DSC.

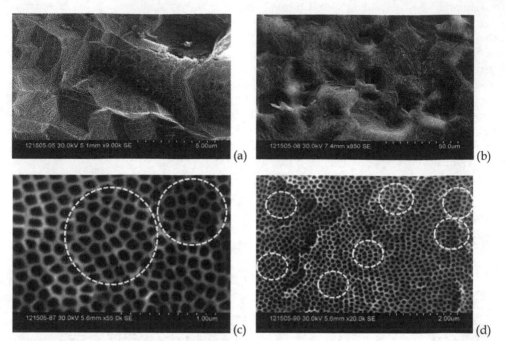

Fig. 10. SEM images of TiO2 NTs under 5 hrs anodization on etched Ti substrate.
(a) anodized surface at higher magnification; **(b)** anodized surface at lower magnification;
(c) highly ordered NT with local concavity shown highlighted in yellow circle; **(d)** several
concave geometries highlighted in yellow circles

4. Effect of TiO$_2$ NT morphology on PV performance of DSC

The TiO$_2$ NTs on the Ti substrate can be used directly for the fabrication of DSCs. The Ti
metal will function as same as TCO layer in conventional Gratzel type DSCs. Due to the
non-transparency of Ti metal to the sunlight, the cell has to be illuminated from counter
electrode (back illumination). In 2007, Grimes *et al* reported 6.89 % conversion efficiency of
this type of cell using ruthenium dye (N719) as light absorber, 20 μm long TiO$_2$ NT arrays
for dye adsorption, and iodide/tri-iodide as electrolyte. Several other groups who fabricated
DSCs with this configuration achieved efficiencies ~3% under similar conditions. In 2009,
Grätzel *et al* reported a 3.59% conversion efficiency of DSCs using ruthenium dye (N719) as
a light absorber, 14 μm long TiO$_2$ NT array for dye adsorption, and ionic liquid as
electrolyte. He *et al* also achieved an efficiency of 3.45% with this configuration. Since TiO$_2$
NT arrays are often attached on the Ti foil and difficult to lift off, the NT arrays with Ti foil
were used directly for cell fabrication. Sunlight must come from the rear of the cell. The
absorption and reflection of sunlight by electrolyte and Pt counterelectrode respectively lead
to reduction in photon flux reaching the dyes. Various techniques were reported from 2008 –
2010 for the growth, liftoff and fixation of NTs on transparent conducting substrate but they
either lacked reproducibility or was time consuming.

In our work Freestanding NT films were obtained by preferential etching of the TiO_2/Ti interface followed by its fixation on TCO with colloidal TiO_2 paste as adhesive layer. The SEM image of freestanding NT film reveals that one end of the NT is open while other end is closed rendering it to be like a test tube structure. Figure 11 (a) shows the morphology of open end of NT while Figure 11 (b) shows the surface morphology of the closed end side of NT. The freestanding NTs can be used in two different orientations for fixation on TCO substrate; one with open end of NT facing the substrate while other with closed end of NT facing the substrate as shown in Figure 12. This section thus tends to highlight the effect of NT orientation on DSC performance. It was reported earlier that the closed end of NT facing the substrate might be helpful in improving the efficiency of DSC by serving as a barrier layer in between substrate and TiO_2 active layer improving the charge transport by minimizing the substrate/TiO_2 interface recombination analogous to the compact layer in NP based DSC.

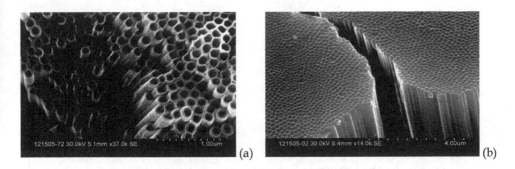

Fig. 11. (a) Top view of NT showing one end to be open; (b) bottom view of NT showing other end to be closed

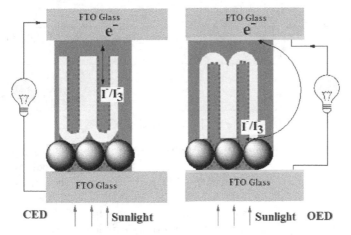

Fig. 12. Simplified DSC structure with CED and OED orientation of NT on TCO

In contrast it was also reported that ~ 2 – 3 μm thick layer of TiO$_2$ at the closed end of NT might serve as an insulating layer between TCO/TiO$_2$ layers which can be detrimental for effective charge transport from active layer to the electrode. In order to investigate the effect of closed end layer on PV performance, we fabricated DSC with two different orientations i.e. closed end facing the substrate and open end facing the substrate, hereafter referred to as CED and OED respectively. The DSCs fabricated with these two structures have apparently shown a big difference in their PV performance as can be seen from the J-V characteristics shown in Figure 13 (a).

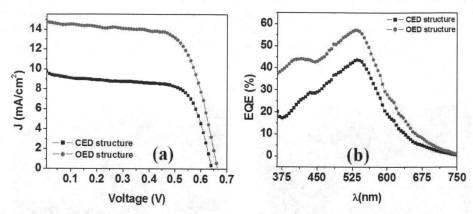

Fig. 13. **(a)** J-V curve under illumination for cells with OED and CED structures; **(b)** EQE curves for cells with OED and CED structures

It was found that the OED structure had higher efficiency of 6.58% as opposed to 4.17% efficiency of CED structure. It was found that cell with OED structure exhibited higher values of short circuit current density (J$_{SC}$), open circuit voltage (V$_{OC}$) and fill factor (FF) compared to CED structure. The J-V data for the photovoltaic performance of two cells is provided in Table 1.

Orientation of NT	NP layer thickness (μm)	NT length (μm)	J$_{SC}$ (mA/cm^2)	V$_{OC}$ (mV)	FF (%)	η (%)
OED	3	22	14.75	666	67.05	6.58
CED	2.7	23	9.5	642	68.45	4.17

Table 1. J-V data for cells with OED and CED orientation of NTs.

In order to further support our J-V data we performed the external quantum efficiency (EQE) measurements on two cells as shown in Figure 13 (b). The EQE data was found to be very consistent with our J-V data where OED structure have shown greater quantum efficiency compared to CED structure. The current densities calculated from the EQE measurements were found to be ~ 15 and 10 mA/cm^2 for OED and CED structures respectively which were in close agreement with the J-V data. Overall the cell performance indicated the superiority of the OED over CED orientation.

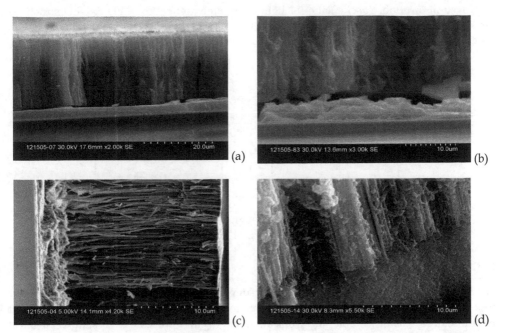

Fig. 14. Shows the cross-sectional SEM image of TiO2 NTs on FTO glass (**a**) CED orientation; (**b**) CED NT/TiO2 NP interface; (**c**) OED orientation; (**d**) OED NT/ TiO2 NP interface

In order to investigate the reason for difference in the PV performance of two structures we performed the cross-sectional SEM imaging of CED and OED structures shown in Figure 14. The interface between colloidal TiO$_2$ NP layer and the NT for CED structure (shown in Figure 14 (a & b)) can be seen to have gaps in between these two layers which suggest that the electron transfer between these two layers is not efficient leading to excessive slow down of the electrons at this interface increasing the recombination probability. We attribute the poor interface quality of this structure to the round shaped closed end of the NT which might have prevented the colloidal particles to partially penetrate into the tube leading to weak interface formation which upon high temperature sintering of the film might have introduced gaps at the interface. Interestingly this feature was not observed in the case of OED structure as can be seen from the cross-sectional image of Figure 14 (c & d). The NTs were found to have formed very good interface by embedding itself into the NP matrix leaving behind no gaps. It can be seen from the image that even after sintering at high temperature the interface retained its good morphology.

In order to investigate the reason for higher photocurrent in OED structure we performed dye loading measurements for two cells. The dye loading densities for cells with OED and CED structures were found to be ~ 7.16 x 10^{-6} mol g^{-1} and 3.58 x 10^{-6} mol g^{-1} respectively which indicates higher dye loading for OED compared to CED structure and hence higher photocurrent. In addition we also anticipate that the improved photocurrent can also be a result of higher confinement of light in the active layer of TiO$_2$ due to the nano-dome structure of closed end being on top leading to the increase in optical path length and hence improved absorption. A schematic for light confinement effect for CED & OED structures are shown in Figure 15 (a & b) respectively. Overall it can be seen that orientation of the NTs for cell fabrication also plays a critical role in determining the efficiency of DSC.

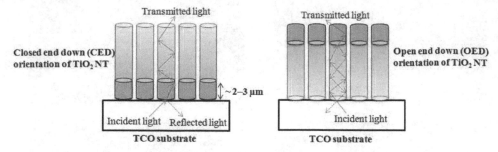

Fig. 15. (a) Schematic of light propagation through NT photoelectrode on FTO with (a) CED structure; and (b) OED structure

5. Conclusions

We found that morphology of NTs largely depends on the macro and microstructural topology of the substrate. Removal of substrate induced disorder in the morphology might be difficult by using simple ultrasonication or drying processes. A simple chemical pretreatment of substrate leads to substantial change in the morphology of grown NTs that can help in obtaining highly oriented and ordered TiO$_2$ NT arrays. The chemical pretreatment technique can find potential utility for being simple, cost effective and less time consuming. In addition we also found that the orientation of the NTs was critical in determining the efficiency of DSC. Hence a meticulous choice of NT orientation along with surface texturing of substrate can significantly help in engineering NT morphology for its successful implementation as a promising material for solar cells as well as other optoelectronic device applications.

6. References

Albu, S.P.; Ghicov, A., Macak, J.M., Hahn, R. & Schmuki, P. (2007). Self-organized , free-standing TiO$_2$ NT membrane for flow through photocatalytic applications. (April, 2007). Nano Letts., 7, 5, 1286-1289

Ali, G.; Chen, C., Yoo, S.H., Kum, J.M. & Cho, S.O. (2011). Fabrication of complete titania nanoporous structures via electrochemical anodization of Ti. *Nanoscale Research Letts.* 2011, 6:332

Anta, J.A.; Casanueva, F. & Oskam, G. (2006). A numerical model for charge transport and recombination in dye-sensitized solar cells. (March, 2006). *J. Phys. Chem. B.*, 110, 5372-5378

Bisquert, J. & Vikhrenko, V. S. (2004). Interpretation of the time constant measured by kinetic techniques in nanostructured semiconductor electrodes and dye-sensitized solar cells. (January, 2004). *J. Phys. Chem. B.*, 108, 2313-2322

Bisquert, J. & Zaban, A (2003). The trap limited diffusivity of electrons in nanoporous semiconductor networks permeated with a conductive phase (November, 2002). *Appl. Phys. A.*, 77, 507-514

Burke, A.; Ito, S, Snaith, H., Bach, U., Kwiatkowski, J. & Grätzel, M (2008). The function of a TiO compact layer in dye-sensitized solar cells incorporating "planar" organic dyes (March, 2008). *Nano Letts.* 8, 4, 976-981

Cameron, P.J. & Peter, L.M. (2003). Characterization of titanium dioxide blocking layers in dye-sensitized nanocrystalline solar cells (October, 2003). *J. Phys. Chem. B.*, 107, 14394-14400

Cameron, P.J. & Peter, L.M. (2005). How important is the back reaction of electrons via the substrate in dye-sensitized nanocrystalline solar cells? (October, 2004). *J. Phys. Chem. B.*, 109, 930-936

Cao, C.; Zhang, G., Song, X. & Sun, Z. (2011). Morphology and microstructure of as synthesized anodic TiO₂ NT arrays. *Nanoscale Res. Lett* (2011), 6, 64

Cao, F.; Oskam, G., Meyer, G.J. & Searson, P.C. (1996). Electron transport in porous nanocrystalline TiO₂ photoelectrochemical cells (October, 1996). *J. Phys. Chem*, 100, 17021-17027

Cass, M. J.; Qiu, F.L., Walker, A.B., Fisher, A.C. & Peter, L.M. (2003). Influence of grain morphology on electron transport in dye sensitized nanocrystalline solar cells (October, 2002). *J. Phys. Chem. B*, 107, 113-119

Chen, Q. & Xu, D. (2009). Large-scale, noncurling, and free-standing crystallized TiO₂ NT arrays for dye-sensitized solar cells. (March, 2009). *J. Phys. Chem. C.*, 113, 15, 6310-6314

Chen, Q.; Xu, D., Wu, Z. & Liu, Z. (2008). Free-standing TiO₂ NT arrays made by anodic oxidation and ultrasonic splitting. (July, 2008). *Nanotechnology*, 19, 365708

Crawford, G.A. & Chawla, N. (2009). Tailoring TiO₂ NT growth during anodic oxidation by crystallographic orientation of Ti. (February, 2009)., *Scripta Materialia*, 60, 874-877

Dor, S.; Grinis, L., Ruhle, S. & Zaban. A. (2009). Electrochemistry in mesoporous electrodes: Influence of nanoporosity on the chemical potential of the electrolyte in dye sensitized solar cells (January 2009). *J. Phys. Chem. C*, 113 (5), 2022-2027

Dubey, M. & He, H. (2009). Morphological studies of vertically aligned TiO₂ NTs for high efficiency solar cell. *Proc. 34th IEEE Photovolt. Conf.* (2009), 002130

Dubey, M.; Shrestha, M., Zhong, Y., Galipeau, D. & He, H. (2011). TiO₂ NT membranes on transparent conducting glass for high efficiency dye-sensitized solar cells. (May, 2011)., *Nanotechnology*, 22, 285201

Ghicov, A. & Schmuki, P. (2009). Self-ordering electrochemistry: a review on growth and functionality of TiO₂ NTs and other self aligned MOₓ structures. *Chem. Commun.* (April, 2009), 2791-2808

Han, L.; Koide, N., Chiba, Y., Islam, A., Komiya, R., Fuke, N., Fukui, A. & Yamanaka, R. (2005). Improvement of efficiency of dye-sensitized solar cells by reduction of internal resistance. *Appl. Phys. Letts.* (May, 2005), 86, 213501

He, H.; Dubey, M., Zhong, Y., Shrestha, M. & Sykes, A.G. (2011). 2-(1-Acetyl-2-oxopropyl)-5,10,15,20-tetraphenyl Porphyrin and its transition metal complexes. (July, 2011). *Eur. J. Inorg. Chem.* 25, 3731-3738

He, H.; Sykes, A.G., Dubey, M., Yan, X., Galipeau, D. & Ropp, M. (2008). *Proc. 33rd IEEE Photovolt. Spec. Conf.*

In, S-I.; Hou, Y., Abrams, B.L., Vesborg, P.C.K. & Chorkendorff, I. (2010). Controlled directional growth of TiO₂ NTs. (March, 2010). *Journal of The Electrochemical Society*, 157, (5), E69-E74

Jennings, J. R.; Ghicov, A., Peter, L.M., Schmuki, P. & Walker, A.B. (2008). Dye-sensitized solar cells based on oriented TiO NT arrays: transport, trapping and transfer of electrons. (September, 2008). *J. Am. Chem. Soc.*, 130, 40, 13364-13372

Kang, S.H.; Kim, H.S., Kim, J-Y. & Sung, Y-E. (2009). An investigation on electron behavior employing vertically-aligned TiO₂ NT electrodes for dye-sensitized solar cells. (August, 2009), *Nanotechnology*, 20, 355307

Kang, S.H.; Kim, J-Y., Kim, H-S. & Sung, Y-E. (2008). Formation and mechanistic study of self-ordered TiO₂ NTs on Ti substrate. (June, 2007). *Journal of Industrial and Engineering Chemistry*, 14, 52-59

Karthikeyan, C.S. & Thelakkat, M (2008). Key aspects of individual layers in solid-state dye-sensitized solar cells and novel concepts to improve their performance (April, 2007). *Inorganica Chimica Acta*, 361, 635-655

Kontos, A.G.; Kontos, A.I., Tsoukleirs, D.S., Likodimos, V., Kunze, J., Schmuki, P. & Falaras, P. (2009). Photo-induced effects on self organized TiO₂ NT arrays: the influence of surface morphology. (December, 2008). *Nanotechnology*, 20, 045603

Lee, K-M.; Suryanarayanan, V. & Ho, K-C (2006). The influence of surface morphology of TiO₂ coating on the performance of dye-sensitized solar cells (May, 2006). *Solar Energy materials & Solar Cells*, 90, 2398-2404

Li, S.; Zhang, G, Guo, D, Yu, L & Zhang, W. (2009). Anodization fabrication of highly ordered TiO₂ nanotubes. (May, 2009), J. Phys. Chem. C, 113, 12759-12765

Liberator, M.; Burtone, L., Brown, T.M.., Reale, A., Carlo, A.D., Decker, F., Caramori, S. & Bignozzi, C.A. (2009). On the effect of Al₂O₃ blocking layer on the performance of dye solar cells with cobalt based electrolytes. (April, 2009). *Appl. Phys. Lett.* 94, 173113

Lin, C.J.; Yu, W-Y. & Chien, S-H. (2010). Transparent-electrodes of ordered opened-ended TiO₂-NT arrays for highly efficient dye-sensitized solar cells. (December, 2009). *J. Mater. Chem.*, 20, 1073-1077

Macak, J.M.; Hildebrand , H., Marten-Jhans, U. & Schmuki, P. (2008). Mechanistic aspects and growth of large diameter self-organized TiO₂ NTs. (January, 2008). *Journal of Electroanalytical Chemistry.*, 621, 254-266

Mohammadpour, A. & Shankar, K. (2010) Anodic TiO₂ NT arrays with optical wavelength-sized apertures. *J. Mater. Chem.* (September, 2010), 20, 8474-8477

Mor, G. K.; Varghese, O.K., Paulose, M. & Grimes, C.A. (2005). Transparent highly ordered TiO₂ NT arrays via anodization of titanium thin films. *Adv. Funct. Mater.*, 15, 1291-1296

Nazeeruddin, M. K.; Angelis, F. D., Fantacci, S., Selloni, A., Viscardi, G., Liska, P., Ito, S., Takeru, B. & Grätzel. M. (2005). Combined experimental and DFT-TDDFT computational study of photoelectrochemical cell ruthenium sensitizers. (November, 2005). *J. Am. Chem. Soc.*, 127, 48, 16835-16847

Nusbaumer, H.; Zakeeruddin, S.M., Moser, J-E. & Grätzel, M (2003). An alternative redox couple for the dye-sensitized solar cell system. *Chem. Eur. J.* 2003, 9, 3756-3763

O'Regan, B. & Grätzel, M (1991). A low-cost, high-efficiency solar cell based on dye-sensitized colloidal TiO₂ films. (October, 1991). *Nature*, 353, 737-740

Ofir, A.; Grinis, L. & Zaban, A. (2008). Direct measurement of the recombination losses via the transparent conductive substrate in dye sensitized solar cells. (January, 2008). *J. Phys. Chem. C.*, 112, 2279-2783

Oshaki, Y.; Masaki, N., Kitamura, T., Wada, Y., Okamoto, T., Sekino, T., Niihara, K. & Yanagida, S. (2005). Dye-sensitized TiO₂ NT solar cells: fabrication and electronic characterization. (October, 2005). *Phys. Chem. Chem. Phys.*, 7, 4157-4163

Park, J. H.; Lee, T-W. & Kang, M.G. (2008). Growth, detachment and transfer of highly-ordered TiO₂ NT arrays: use in dye-sensitized solar cells. (May, 2008). *Chem. Commun.*, 2867-2869

Prakasam, H. E.; Shankar, K., Paulose, M., Varghese, O.K. & Grimes, C.A. (2007). A new benchmark for TiO₂ NT array growth by anodization. (April, 2007). *J. Phys. Chem. C.*, 111, 20, 7235-7241

Robertson, N. (2006). Optimizing dyes for dye-sensitized solar cells (2006). *Angew. Chem. Int. Ed.* 45, 2338-2345

Roy, P.; Kim, D., Lee, K., Spiecker, E. & Schmuki, P. (2010). TiO₂ NTs and their application in dye-sensitized solar cells. *Nanoscale.* (December, 2009), 2, 45-49

Ruan, C.; Paulose, M., Varghese, O. K., Mor, G. K. & Grimes, C.A. (2005). Fabrication of highly ordered TiO₂ NT arrays using an organic electrolyte. (July, 2005)., *J. Phys. Chem. B.*, 109, 33, 15754-15759

Santiago, F.; Bisquert, J., Belmonte, G., Boschloo, G. & Hagfeldt, A. (2005). Influence of electrolyte in transport and recombination in dye-sensitized solar cells studied by impedance spectroscopy (November, 2004). *Solar Energy Material & Solar Cells*, 87, 117-131

Santiago, F.F.; Barea, E.M., Bisquert, J., Mor, G.K., Shankar, K. & Grimes, C.A. (2008). High carrier density and capacitance in TiO NT arrays induced by electrochemical doping. (August, 2008). *J. Am. Chem. Soc.*, 130, 34, 11312-11316

Santiago, F.F.; Belmonte, G.G., Bisquert, J., Zaban, A. & Salvador, P. (2002). Decoupling of transport, charge storage, and interfacial charge transfer in the nanocrystalline TiO₂/electrolyte system by impedance methods. (December, 2001). *J. Phys. Chem. B.*, 106, 334-339

Santiago, F.F.; Bisquert, J., Belmonte, G.G., Boschloo, G. & Hagfeldt, A. (2005). Influence of electrolyte in transport and recombination in dye-sensitized solar cells studied by impedance spectroscopy (November, 2004). *Solar Energy Materials & Solar Cells*, 87, 117-131

Sero, I. M.; Dittrich, T., Belaidi, A., Belmonte, G.G. & Bisquert, J. (2005). Observation of diffusion and tunneling recombination of dye-photoinjected electrons in ultrathin TiO₂ layers by surface photovoltage transients. (May, 2005). *J. Phys. Chem. B.* 109, 14932-14938

Sero, I.M. & Bisquert, J. (2003). Fermi level of surface states in TiO₂ NPs. (June, 2003). *Nano Letts.* 3, 7, 945-949

Sero, I.M.; Dittrich, T., Belmonte, G.G. & Bisquert, J. (2006). Determination of spatial charge separation of diffusing electrons by transient photovoltage measurements. (August, 2006). *J. App. Phys.* 100, 1

Shiga, A.; Tsujiko, A., Ide, T., Yae, S. & Nakato, Y (1998). Nature of electrical junction at the TiO₂/substrate interface for particulate TiO₂ film electrodes in aqueous electrolyte (May, 1998). *J. Phys. Chem. B.*, 102, 6049-6055

226

Shin, Y. & Lee, S. (2008). Self-organized regular arrays of anodic TiO$_2$ nanotubes. (September, 2008), Nano Letts. 8, 10, 3171-3173

Sun, L.; Zhang, S., Sun, X. & He, X. (2010). Effect of the geometry of the anodized titania NT array on the performance of dye-sensitized solar cells. *J. Nanosci. Nanotechnol.* , 10, 1-10

Tschirch, J.; Bahnemann, D., Wark, M. & Rathousky, J. (2008). A comparative study into the photocatalytic properties of thin mesoporous layers of TiO$_2$ with controlled mesoporosity (August, 2007). *Journal of Photochemistry and Photobiology A: Chemistry*, 194, 181-188

Vanmaekelbergh, D. & de Jongh, P.E (1999). Driving force for electron transport in porous nanostructured photoelectrode (January, 1999). *J. Phys. Chem. B.*, 103, 5, 747-750

Varghese, O.K.; Paulose, M. & Grimes, C.A. (2010). Long vertically aligned titania NTs on transparent conducting oxide for highly efficient solar cells. (August, 2009). *N. Nano.2009.226*

Wang, J. & Lin. Z. (2008). Freestanding NT arrays with ultrahigh aspect ratio via electrochemical anodization. (November, 2007). *Chem. Mater.* 20, 1257-1261

Wei, M.; Konishi, Y., Zhou, H., Yanagida, M., Sugihara, H. & Arakawa, H. (2006). Highly efficient dye-sensitized solar cells composed of mesoporous titanium dioxide (January, 2006). *J. Mater. Chem.*, 16, 1287-1293

Xu, T.; He, H., Wang, Q., Dubey, M., Galipeau, D. & Ropp, M. (2008). *Proc. 33rd IEEE Photovolt. Spec. Conf.*

Yuan, L.; Xurui, X., Dongshe, Z., Puhui, X. & Baowen, Z (2003). Light scattering characteristic of TiO$_2$ nanocrystalline porous films (May, 2003). *Chinese Science Bulletin*, 48, 9

Zhang, L. & Han, Y. (2010). Effect of nanostructured titanium on anodization growth of self-organized TiO$_2$ nanotubes. (December, 2009), Nanotechnology., 21, 055602

Zhu, J.; Hsu, C-M., Yu, Z., Fan, S. & Cui, Y. (2010). Nanodome solar cells with efficient light management and self-cleaning. (November, 2009). *Nano Lett.* , 10, 1979-1984

Zhu, K.; Neale, N.R., Miedaner, A. & Frank, A. J. (2007). Enhanced charge collection efficiencies and light scattering in dye-sensitized solar cells using oriented TiO$_2$ NT arrays. (December, 2006), *Nano Letts.*, 7, 1, 69-74

Zhu, K.; Vinzant, T.B., Neale, N.R. & Frank, A.J. (2007). Removing structural disorder from oriented TiO NT arrays: Reducing the dimensionality of transport and recombination in dye-sensitized solar cells. (November, 2007). *Nano Lett.*, 7, 12, 3739-3746

12

Exploring the Superconductors with Scanning Electron Microscopy (SEM)

Shiva Kumar Singh[1,2,*], Devina Sharma[1], M. Husain[2],
H. Kishan[1], Ranjan Kumar[3] and V.P.S. Awana[1]

[1]*Quantum Phenomena and Applications,*
National Physical Laboratory (CSIR), New Delhi,
[2]*Department of Physics, Jamia Millia Islamia, New Delhi,*
[3]*Department of Physics, Panjab University, Chandigrah,*
India

1. Introduction

The characterization of materials supports their development and in particular of superconductors, for their technological applications. Scanning electron microscopy (SEM) is one of these characterization techniques, whose data is used to estimate the properties, determine the shortcomings and hence improve the material. The phenomenon of superconductivity initially develops within the grain and eventually crosses over the grain boundaries, leading to the bulk. Hence SEM can be a useful tool to probe the microstructure of the superconductors and the properties related to it. Along with this the Energy-dispersive Spectroscopy (EDS) can tell about the chemical composition of compounds. Grain size and its connectivity can be seen through SEM and can be correlated with the corresponding properties. The superconducting materials developed for practical applications are some of the complex materials used today. These materials have large number of potential variables such as their processing conditions, composition, structure etc., whose dependence on the superconducting properties have to be analyzed critically. The characterization techniques are the tools that help to reveal and explore both the macro and microstructure of materials. It is known that the larger grains (reduction in grain boundaries) lead to increased pinning type behavior with enhanced J_c [1]. In contrast Rosko et al. [2] reported that J_c is determined by weak links and grain size has little role on it. Also, Smith et al. [3] interpreted reduction of J_c and activation of weak link type behavior with increasing grain size for $YBa_2Cu_3O_{7-\delta}$ (YBCO) polycrystalline samples in terms of micro-cracks in large grains. The superconducting parameters are broadly divided into two categories; first, the intrinsic parameters such as penetration depth (λ), which are intrinsic to the material and are not affected by, grain size. On the other hand, values such as shielding/Meissner fraction, the inter- and intra-grain critical current density and diamagnetic fraction depend upon particle size of bulk superconductors. Thus SEM can be very important to probe and in understanding the superconducting phenomena.

---

* Corresponding Author

In SEM electron beam is scanned across a sample's surface. When the electrons strike the sample, a verity of signals arises and produces elemental composition of the sample. *SEM* with *EDS* is a major tool for qualitative and quantitative analyses which is done by bombarding a finely focused electron beam (electron probe) on the specimen, and measuring the intensities of the characteristic X-ray emitted. The three signals in SEM are the secondary electrons, backscattered electrons and X-rays, provide the greatest amounts of information. Secondary electrons are emitted from the atoms occupying the top surface and produce interpretable image of the surface. The contrast in the image is determined by the sample morphology. Backscattered electrons are primary electrons which are "reflected" from atoms in the solid. The contrast in the image produced is determinate by the atomic numbers of the elements in the sample. Therefore the image shows the distribution of different chemical elements in the sample. Since these electrons are emitted from the depth of the sample, the resolution of the image is not as good as for secondary electrons.

This chapter deals with the ability of SEM in extracting the information from superconductors. Since, the microstructure and topology of the materials determine largely its properties in terms of its grains and their connectivity, the utility of SEM in studying the properties of various superconductors discovered till date will be reviewed. The limitations will also be discussed. How other characterization tools, can provide better information along with *SEM*, will be explored.

2. SEM: As a characterization tool for superconductors

Some of the important aspects with which *SEM* deals with, is the grain size, morphology and alignment, structural defects, chemical composition. While trying to optimize the transport properties, grain to grain alignment within the superconductor has to be considered. It is important to analyze the grain alignment and enhance it in order to relate it to the improvement of transport properties. Defects play an important role in determining the properties of superconductors, especially those of HTSc. While macro defects such as porosity, cracks, secondary phases etc. may adversely affect the transport critical super currents, on the other hand microscopic defects such as addition of nanoparticles, dislocation etc., can prove to be beneficial. As the size of the defects is smaller than the coherence length of the superconductor, they may act as pinning centers, thereby enhancing the critical current. Chemical composition of the material within the superconducting grain and at the grain boundaries has significant effect over its properties. The compositions and the change in the compositions can be measured or inferred from a variety of techniques such as energy dispersive spectroscopy. Backscattered electron imaging (BEI) can also be used to infer chemical compositional variations. In all, the complexity in the superconducting materials requires continuous research into their fundamental properties and evolution of new improved materials. For all these evaluations, various characterization tools have to be relied upon among which SEM and allied techniques such as EDS and BEI play an important role.

2.1 Determination of grain size from *SEM* micrographs

Grain size determination is perhaps one of the most commonly performed microstructural measurements from *SEM* micrographs. Standards organizations, including American Society for Testing and Materials (ASTM) [4] and some other national and international organizations, have developed standard test methods describing how to characterize microstructures quantitatively. The methods for grain size measurement are described in great detail in the ASTM Standard, E112, "Standard Test Methods for Determining Average Grain Size" [5]. The information below will provide cursory explanation to the methods for determining grain size in *ASTM* E112. The microstructural quantity known as the *ASTM* micro gain size number, G is defined as

$$n = 2^{G-1}$$

Where, n is the number of grains per square inch measured at a magnification of 100x. Grain size or the value of G is most commonly measured by (a) Planimetric method and by (b) Intercept method.

(a) Planimetric method

In the Planimetric method [Fig. 1 (a)] (developed by Zay Jeffries in 1916), a count is made of the number of grains completely within a circle of known area and half of the number of grains intersected by the circle to obtain N_A. Then, N_A is related to G. This method is slow when done manually because the grains must be marked when counted to obtain an accurate count. This method is described in the section 9 of *ASTM E112*.

The basic steps of the procedure are as follows:

i.   A circle of known size is inscribed over the SEM image.
ii.   The numbers of grains are counted that are completely within the area (n_{inside}).
iii. The number of grains is counted that are partially within the area ($n_{intercepted}$).
iv.   The number of grains per sq. mm, N_A, is calculated from $N_A = f \{n_{inside} + \frac{1}{2}(n_{intercepted})\}$
v.   The multiplier f is calculated from (M^2/circle area), where M is the linear magnification of the image.
vi.   From N_A, we can calculate the *ASTM* grain size number, G, using the following formula from E 112-96: $G = \{3.322 (\log 10 \ N_A) - 2.954\}$

(b) Linear intercept method

The *ASTM* grain size can also be determined using the intercept method (developed by Emil Heyn in 1904) counting either the number of grains intercepted (N) or the number of grain boundaries intersected (P) with a test line. *ASTM* recommends using a grid with three concentric circles (as shown in the Fig. 1 (b)) with a 500mm total line length. The count of the number of grains intercepted by the circle is N. To calculate the number of interceptions per mm, N_L, we divide N by the true length of the circle. The true length (L_T) is obtained by dividing the circumference of the circle by the magnification, M. Hence, $N_L = N/L_T$ interceptions per mm. To calculate the grain size, we first determine the mean linear intercept length, l, which is the reciprocal of N_L (or of P_L, the number of grain boundary intersections per unit length). G is calculated from an equation from E 112-96:

$$G = \{-6.644 (\log 10 \ l) - 3.288\},$$

where, l is in mm.

ASTM E112 provides table that relates grains/in^2 @ 100x and grains/mm^2 @ 1x to *ASTM* grain size number G. Since the two methods are sensing different geometric aspects of the three-dimensional grain structure, they will not give exactly the same value, but they will be close, generally within the experimental limitations of the measurements. In practice, these measurements are repeated on a number of fields in order to obtain a good estimate of the grain size.

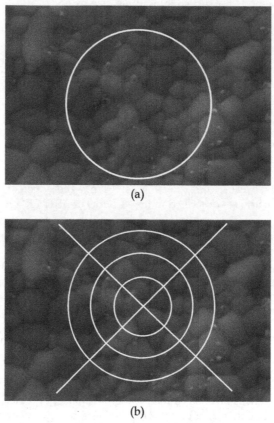

Fig. 1. **(a)** Planimetric method, **(b)** Linear Intercept method

2.2 Chemical composition using EDS

One of the most outstanding features of the *SEM-EDS* is that it allows elemental analysis and observation from an ultra micro area to a wide area on the specimen surface without destroying the specimen. Qualitative and quantitative analysis (by *EDS*) by electron probe takes advantage of the emission of characteristic X-radiation by electron interactions in the valence shell of atoms. Backscattered electron images in the *SEM* display compositional contrast that results from different atomic number and their distribution of elements. *EDS* allows one to identify what those particular elements are and their relative proportions, for example their atomic percentage. Initial EDS analysis usually involves the generation of an X-ray spectrum from the entire scan area of the SEM. In the X ray spectra generated from the entire scan, Y-axis shows the counts (number of X-rays received and processed by the detector) and the X-axis shows the energy level of those counts. The EDS software associates the energy level of the X-rays with the elements and shell levels that generated them.

3. Exploring superconductors using *SEM*

In this section we will discuss the *SEM* study of superconductors of different families. For high temperature superconductors (HTSc), *SEM* has been widely used to explore the superconducting behavior. Grain size matters a lot in deciding the superconducting parameters of cuprate HTSc. Decrease in shielding current is observed with decrease in particle size [6]. Magnesium diboride (MgB_2) represent an attractive alternative to low temperature superconductors. For most of the practical applications, high critical current density (J_c) in the presence of a magnetic field along with high upper critical field (H_{C2}) and high irreversibility field (H_{irr}) are required. Moderate impurities and nano (n) materials are being used to improve these parameters [7-14]. In particular, significant flux pinning enhancement in MgB_2 is observed with n-SiC addition [15]. In case of newly discovered pnictides superconductors, *SEM* is being used in an ingenious way. The spatially resolved electrical transport properties have been studied on the surface of optimally-doped superconducting $Ba(Fe_{1-x}Co_x)_2As_2$ single crystal by using a four-probe scanning tunneling microscopy [16]. Results will be discussed in following subtitles.

3.1 Grain size and gain connectivity in cuprate high temperature superconductors

One of the characteristics of high temperature superconductors (HTSc) is their small coherence length which is comparable with the unit cell. The coherence length is a key parameter for the performance of superconductors for applications, since this determines the size of the normally conducting core of the flux lines [17]. In order to control the motion of flux lines one needs a microstructure with defects as small as the coherence length. The extremely small coherence length of HTSc, which is for YBCO only 2.7 nm at 77 K within the ab-plane, is the reason that defects such as grain boundaries, which are very beneficial in low-temperature superconductors because they act as pinning defects, serve as weak links and limit the critical current, especially in the presence of an external magnetic field. HTSc bulk material can therefore be considered as a matrix of superconducting grains embedded in a non superconducting material.

The cuprate superconductors belong to the family of HTSc in which Cu-O chains and planes are responsible for the conduction of super currents. The size as well as the shape of the grains varies in different cuprate HTSc [18-20]. This variation in microstructure in turn leads to different superconducting behavior. Although the nature of the occurrence of superconductivity in cuprate HTSc is same, T_c varies from 38 K in LSCO to 110 K in BSCCO system. This wide range of T_c itself indicates that micro-structural parameters are of much importance. In this section various cuprate HTSc will be discussed in the chronological order of their discovery, in terms of the use of SEM for their characterization.

A. $La_{2-x}Sr_xCuO_4$

Discovered by Bednorz and Muller in 1986 [21], $La_{2-x}Sr_xCuO_4$ was the trendsetter breakthrough in the history of superconductivity leading to the new era of High T_c superconductivity. It has T_c of 38 K which is beyond the BCS limit of 30 K. Although it has higher T_c than conventional superconductors, its critical parameters such as critical temperature, field and current density, which are applicable in practical applications, are weaker. Various attempts had been made earlier to enhance these parameters and to understand the physics behind it. The values such as shielding/Meissner fraction, the inter- and intra-grain critical current density and diamagnetization fraction depends upon particle size of bulk superconductors. One of the earliest reports on the effect of particle size on the physical properties of a superconductor was by Chiang et al. [22], who varied particle size from 1 to 10 μm and found significant changes in the superconducting and physical properties. Another fact that has well been established is that the critical current measurements done in HTSc have shown a much lower value for polycrystalline bulk samples [1] than single crystals of the same compound. This difference cannot be attributed alone to the intrinsic nature (anisotropy etc.) of the material. In fact, the quasi-insulating grain boundaries of HTSc play a detrimental role in limiting the critical current and other superconducting and magnetic properties [22-23]. In a recent report D. Sharma et al. [18] have investigated the influence of grain size (sintering temperature) on various superconducting parameters in $La_{1.85}Sr_{0.15}CuO_4$. From SEM micrographs [see Fig. 2 (a), (b) and (c)] it is clear that with the increase in the sintering temperature, there is a considerable increase in the grain size. Increment in the grain size brings in double boom to the superconductivity as it reduces the number of insulating grain boundaries (weak links) as well as increases the effective superconducting volume fraction. Thus both the inter and intra critical current density which is limited by the weak links is expected to enhance with increase in the grain size. Qualitative picture given by the SEM micrographs corroborates with the quantitative results obtained from various transport and magnetic measurements as given in the table 1.

Fig. 2. SEM of $La_{1.85}Sr_{0.15}CuO_4$ samples sintered at (a) 900 °C, (b) 1000 °C, and (c) 1050 °C [Ref. 18].

Grain size (2R)	Critical current density J_c (A/cm^2)		Diamagnetization (at $T = 10$ K, $H = 10$ Oe)		Volume fraction not penetrated $(R - \lambda)^3/R^3$
	$J_c = (20 \times \Delta M)/a(1 - a/3b)$ Inter-granular J_c	$J_c = (15 \times \Delta M \times \rho)/R$ Intra-granular J_c	ZFC	FC	
1 μm	1.1×10^3	3.75×10^5	0.36%	0.28%	9.08%
2 μm	3.8×10^3	8.36×10^5	2.10%	1.59%	38.07%
4.5 μm	40.4×10^3	32.73×10^5	16.95%	4.08%	64.20%

Table 1. Critical current density, diamagnetization fraction and percentage volume fraction not penetrated by magnetic flux calculated from magnetic measurement data for different grain sizes of $La_{1.85}Sr_{0.15}CuO_4$ [Ref. 18].

In another particle size controlled study of non-superconducting $La_{1.96}Sr_{0.04}CuO_4$ was made with SEM and IR spectra by S. Zhou et al [24]. They observed that as the particle size reduces, the IR band at around 685 cm-1, corresponding to in-plane Cu–O asymmetrical stretching mode, shifts to higher frequency and the magnetization exhibits a large enhancement at low temperature. A visible spin-glass transition was found under a relatively weak external field in the sample with the largest particle sizes. Whereas the sample with the smallest particle sizes exhibits no visible spin-glass transition. They suggested that surface effects play a dominant role in determining the magnetic properties as the particle size reduces.

B. $YBa_2Cu_3O_{7-\delta}$

Cuprate superconductors are very sensitive to oxygen content. Depending on oxygen content, $YBa_2Cu_3O_{7-\delta}$ crystallizes in two phases. Tetragonal $P4/mmm$ with $\delta = 0.60$ results in non-superconducting and antiferromagnetic YBCO whereas orthorhombic $pmmm$ with $\delta = 0.05$ phase leads to a superconducting YBCO with T_c 93 K [25]. Also, the intra-grain signal depends much on oxygen content of the composition. The structure of YBCO can be viewed as $(Ba,Sr)O/CuO_2/RE/CuO_2/(Ba,Sr)O$ slabs interconnected through a sheet of Cu and O with variable composition of CuO_x. Charge transport and high temperature superconductivity is believed to reside in the CuO_2 planes of all known HTSc cuprates, except that $CuO_{1+\delta}$ chains have been reported to participate in the b-axis transport of $YBa_2Cu_3O_{7-\delta}$ [26]. In $YBa_2Cu_3O_{7-\delta}$ $(CuBa_2YCu_2O_{7-\delta}$, Cu-1212) there are two different Cu sites, namely Cu1 and Cu2. Cu1 resides in CuO_x chains and Cu2 in superconducting CuO_2 planes. Even at macroscopic level, any contravene in integral CuO_2 stacks, affects superconductivity drastically [27-28]. The CuO_x chain acts as a charge reservoir and provides the mobile carriers to superconducting CuO_2 planes.

To understand the physics of the superconducting nature, investigation of doping various elements at Cu1 site was carried by some of us [29]. And it was found that the YBCO structure is versatile and changes with doped elements at Cu1 site. Single phase samples of 1212 type with different MOx layers showed the great flexibility of these rocksalt layers and variable structure formation. With different M, as the oxidation state and ionic state changes, carrier concentration and structure changes as well. While, Nb-, Fe-, Ru- and Al-1212 possess tetragonal $P4/mmm$ space group structure, the Ga-1212 and Co-1212 are crystallized in orthorhombic $Ima2$ space group.

The SEM images [Fig. 3] suggest that with the doping of variety of elements at the Cu1 site in Y-123 structure, the morphology also changes suggesting change in the structure of the new compounds formed. Change in structure was authenticated by the Rietveld analysis also. In another report Nalin *et al.* [30] studied effect of Zn doping at Cu1 site. With ac susceptibility inter and intra granular changes are studied. In the Zn-doped samples, the inter-grain peak got reduced dramatically. In the χ^{ll} plots of Zn doped samples the inter-grain peak superposes with intra-grain peak and inter-grain peak depresses further. It was concluded that as (see *SEM* images [Fig. 4]) the average grain size is increasing with Zn doping. The increased grain size provides more area for the eddy currents loops to persist in the individual grains, thus systematic enhancing the intra-grain peak.

A combined study through *SEM* and *EDS* for the compounds $Y_{1-x}Ca_xBa_{1.9}Nd_{0.1}Cu_3O_y$ (YCBNCO) with $x \leq 0.40$ have been made [31]. Back-scattered electron SEM micrographs of samples with x = 0.10 and 0.30 have been taken. The *SEM* studies proved that the samples with x ≤ 0.20 are homogeneous and stone like grains with typical size of several microns. On the other hand the samples with x ≥ 0.20 are inhomogeneous. The *SEM* micrographs show that the stone-like grains and the sponge-like grains co-exist in the surface of the samples. The *EDS* results show that the constituted elemental ratios in both regions are different and hence the superconducting properties.

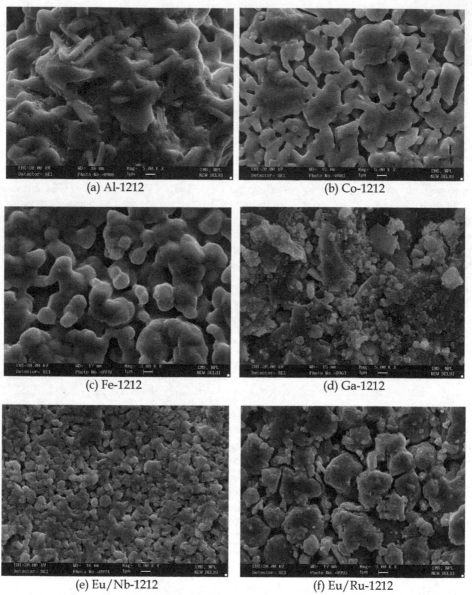

Fig. 3. SEM pictures of the M-1212: (a) AlSr$_2$YCu$_2$O$_{7+\delta}$, (b) CoSr$_2$YCu$_2$O$_7$, (c) FeSr$_2$YCu$_2$O$_{7+\delta}$, (d) GaSr$_2$YCu$_2$O$_7$, (e) NbSr$_2$EuCu$_2$O$_{7+\delta}$ and (f) RuSr$_2$EuCu$_2$O$_{7+\delta}$. [Ref. 29].

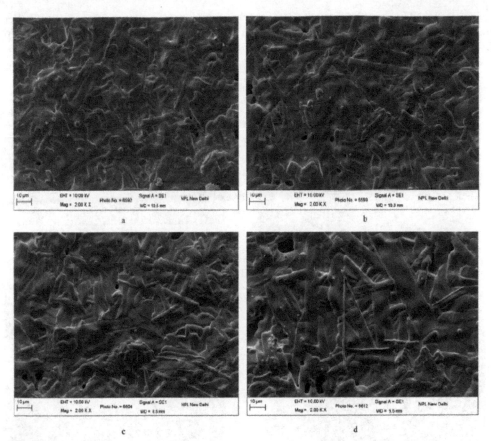

Fig. 4. SEM images of $YBa_2Cu_{3-x}Zn_xO_{7-\delta}$, (a) $x = 0.01$, (b) $x = 0.03$, (c) $x = 0.05$ and (d) $x = 0.10$ [Ref. 30].

C. $(Bi, Pb)_2Sr_2Ca_{n-1}Cu_nOy$

Bismuth based superconducting cuprates (in short named as BSCCO) is an another family of cuprate HTSc which are expressed by a general formula of $(Bi, Pb)_2Sr_2Ca_{n-1}Cu_nOy$. For ($n = 1, 2, 3$), these are abbreviated as Bi2201, Bi2212 and Bi2223 phases, whose superconducting transition temperatures (T_c) are around 20, 85 and 110 K respectively [32]. Though the mechanism of superconductivity in HTSc superconductors has been extensively studied, it is still unclear. As a result of substitution experiments it is well known, that in HTSc's, there is a strong relationship between carrier concentration and transition temperature. In addition, intergrain carrier transportation is also a key factor in deciding the sharpness of the transition. It is a well known fact that intergrain region behaves as a non-conducting region. Thus grain connectivity becomes more important for the sharpness of transition and other critical parameters. The grain growth and its shape varies in $Bi_{2-x}Pb_xSr_2CaCu_2O_8$ (see Fig. 5) with substitution of Pb at Bi site ($0 < x < 0.40$) [19]. From the flake type grain shape in pristine samples to needle type grain shape in $x = 0.40$ composition is observed from SEM micrographs. Moreover, improvement in the packing fraction and hence the inter-granular connectivity was seen in the samples for $x = 0.0$ up to $x = 0.16$, which degrades with further increase in x. Decrease in the grain alignment with increase in Pb content has also been seen in SEM micrographs.

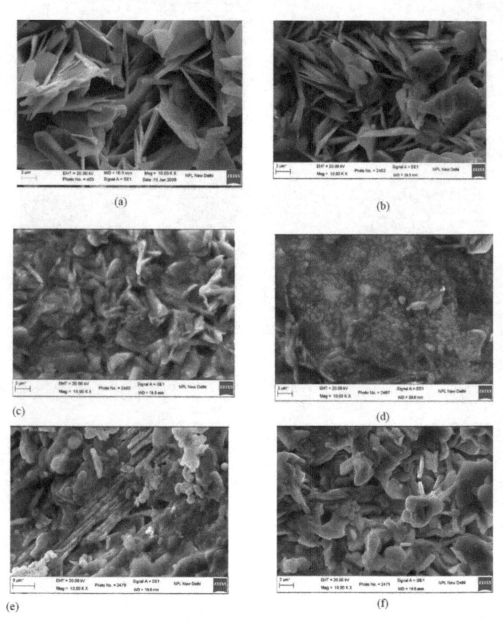

Fig. 5. SEM micrographs of $Bi_{2-x}Pb_x$ $Sr_2CaCu_2O_{8+\delta}$ (a) $x = 0.04$, (b) $x = 0.06$, (c) $x = 0.08$ and (d) $x = 0.16$ (e) $x = 0.20$ and (f) $x = 0.40$[Ref. 19].

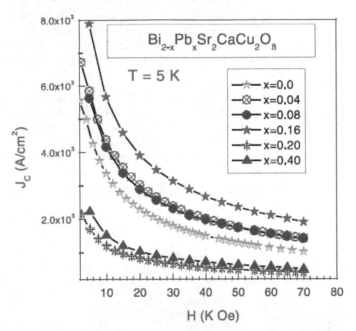

Fig. 6. Jc measurements for $Bi_{2-x}Pb_x Sr_2CaCu_2O_{8+\delta}$ (x = 0 to 0.40) [Ref. 19].

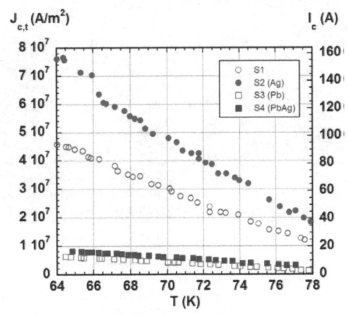

Fig. 7. Temperature dependence of the transport critical current densities, between 65 and 77 K and self-field, for the four samples.
Reprinted with permission from [Supercond. Sci. Technol. 22 (2009) 034012].

The critical current density data (see Fig. 6) for the same samples shows an increase in the current density with Pb substitution in pristine sample till x=0.16 after which it decreases with further increase in x. The decrease in conductivity for samples having $x > 0.16$ has been explained on the basis of the effects arising from decrease of the grain alignment, increase of porosity and secondary phases.

Also, A. Sotelo et al. [20] have studied the Lead (Pb) and Silver (Ag) doping of Bi-2212 samples. It was found that Pb doping results in the decrease of the transport critical current density (see Fig. 7), Jc,t (from 4.4×107 to 6×106 Am−2 at 65 K and self-field) as well as in the worsening of the mechanical properties, by about 35% compared to the undoped samples. In contrast, Ag doping results in the improvement of both the critical current density and mechanical strength.

Growth direction ——————▶

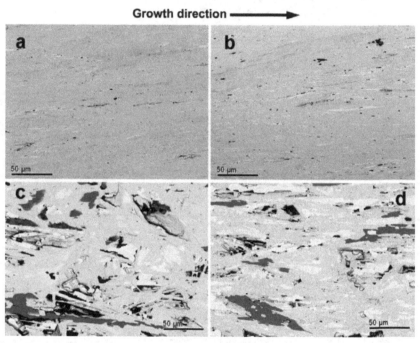

Fig. 8. Longitudinal SEM images obtained on annealed polished samples. (a) S1; (b) S2 (Ag); (c) S3 (Pb); and (d) S4 (PbAg). Phases can be identified as Bi-2201 (white contrast), Bi-free phases (dark grey; CuO, $Sr_{14}Cu_{24}O_{41}$, and $SrCuO_2$), plumbate-like phases and Bi-2212 (grey contrast), Ag (light grey contrast).
Reprinted with permission from [Supercond. Sci. Technol. 22 (2009) 034012].

These described effects are related to the microstructural observations (see Fig. 8) as, Pb doping dramatically reduces the texture, while Ag doping improves it. Moreover, for samples with Ag addition, an intergrowth of Bi-2223 inside the Bi-2212 grains is observed, which would explain the improved superconducting properties of these samples. The stability of these superconductors has been studied through the corrosion process in a moisture atmosphere [33]. By means of optical and SEM observations, several morphologies of the alteration products have been observed.

3.2 Flux pinning in MgB$_2$

Although cuprate superconductors exhibit very high transition temperatures, their in-field performance [34] is compromised by their large anisotropy, the result of which is to restrict high bulk current densities. On the other hand in-field performance (higher Jc), leads diboride of magnesium to much better candidate for application purposes besides its lower T_c than cuprates. With the magneto-optical (MO) and polarized light (PL), SEM was used to assess the issue of inhomogeneous and granular behaviour in MgB$_2$ [35]. It was speculated through SEM that the strongly shielding high-Jc regions are microstructurally subdivided on a scale of 100 nm. Also, in a darker central area a fine mixture of MgB$_2$ and a boron-rich phase was found through SEM [Fig. 9]. They concluded that the strongly shielding regions contain a large number of high-angle grain boundaries. Thus along with MO and PL, SEM suggests that MgB$_2$ is more similar to a low-Tc metallic superconductor than to a high-Tc copper oxide superconductor [35]. A sol-Gel synthesis of MgB$_2$ nanowires is reported Nath et.al [36]. SEM study reveals formation of a thick mesh of nanowires. The nanowires are found to be ca. 50–100 nm in diameter with very smooth surfaces having lengths up to at least 20 micrometer. It is observed that nanowires oriented vertically with respect to the electron beam. Thus SEM also revealed a hexagonal cross section for MgB$_2$ nanowires which is consistent with a degree of crystallinity. The Crystallinity of MgB$_2$ nanowires was also supported by their selected area electron diffraction (SAED) study on some individual nanowires. As MgB$_2$ has better candidature for practical applications various dopants has been added to improve its performance [6-14]. Arpita et al. [7] noted that with n-SiC addition though Tc decreases, but critical current density (Jc) and flux pinning improved significantly. Presence of Mg$_2$Si phase was also revealed through SEM and EDS [Fig. 10]. Dual reaction occurs with n-SiC addition first n-SiC reacts with Mg forming Mg$_2$Si and then free C is incorporated into MgB$_2$ at B site [37]. Thus both reactions help in the pinning of vortices which results in improved superconducting performance. Mg$_2$Si and excess carbon can be embedded within MgB$_2$ grains as nanoinclusions. They argued that due to the substitution of C at the B site the formation of a nanodomain structure takes place due to the variation of Mg-B spacing. These nanodomain defects, having the size of 2-3 nm, can also behave as effective pinning centers. So, highly dispersed nanoinclusions within the grains and the presence of nanodomain defects act as pinning centers and thus result in the improved $Jc(H)$ behavior for the n-SiC doped samples.

3.3 Pnictides: Chemical composition and electrical transport

A. REFeAsO$_{1-x}$F$_x$ (1111)

Iron pnictides are the latest entrant in family of high temperature superconductors [38]. Superconductivity originates in parent pnictides REFeAsO with doping of F at O site. The reactive nature of REs towards oxygen results a very critical synthesis condition for these compounds. Though the compounds are being synthesized in inert/oxygen controlled atmosphere, it is very hard to acquire the desired composition. Thus it is better to analyze the chemical composition of the synthesized compound before going insight and describing the physical properties. Thus SEM with EDS can be very useful in invoking the composition (especially effective F concentration) of the arsenides. The SEM analysis of the parent and non superconducting SmFeAsO compound after metallographic preparation reveals very small amounts of unreacted phases (iron arsenides), which are completely dissolved after

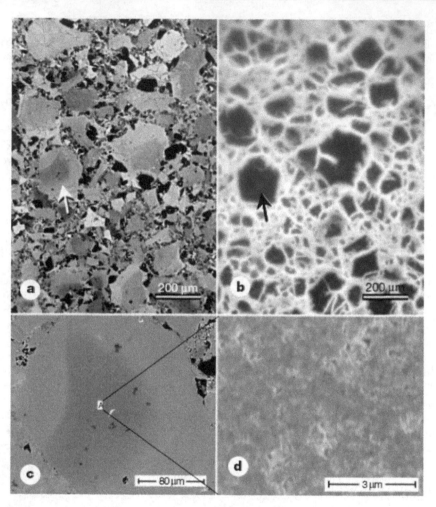

Fig. 9. Polarized light microscope and magneto-optical images of the same area of sample B are compared in (a) and (b), respectively. Bright regions of (b) indicate areas where magnetic flux has penetrated the sample after a field of 120 mT was applied after cooling the sample in zero field to 11 K. Image (c) presents a magnified view using SEM backscattered electron imaging of the strongly superconducting region marked with an arrow in (a) and (b). At higher resolution, image (d), a secondary electron examination of the central region in (c), reveals that the area marked by an arrow in (a) and (b) has, 100-nm, fine-scale structure [Reprinted by permission from Macmillan Publishers Ltd: NATURE 410 (2001) 186].

sintering. In general, sintering greatly increases the density of the samples, but favours the formation of Sm_2O_3 small particles. This feature reveals that at the sintering temperature the formation of Sm_2O_3 competes with the thermodynamic stability of the oxy-pnictide. S. Kaciulis et al. [39] studied $SmFeAsO_{0.85}F_{0.15}$ sample with *SEM, EDS* and *XPS. SEM* image after the fracture manifests the crystals appear clean at the surface, without any contamination of secondary phases. On the other hand another *SEM* image of the same sample after metallographic preparation reveals that crystals are aggregated within a matrix constituted of FeAs, which was also evidenced by their *EDS* analysis. However, their XPS study speculated that the formation of secondary phases, such as FeAs and SmOF. The discrepancy with EDS data, indicating only the presence of FeAs in the matrix, was explained by different analysis depth: up to a few micrometers for *EDS* and only a few nanometers for *XPS*. However, the aggregation of REFeAsO$_{0.85}$ crystals in FeAs matrix is also observed in Back Scattered *SEM* study [Fig. 11] of Ketnami et al. [40]. The absence of significant transport currents in polycrystalline samples has raised the concern that there is a significant depression of the superconducting order parameter at grain boundaries (GB) [41-43]. Remnant magnetization and *MO* studies of polycrystalline NdFeAsO$_{0.85}$ and SmFeAsO$_{0.85}$ uncovered that intergrain and intra-grain current densities had different temperature dependences and differed by three orders of magnitude, leaving open the possibility of an intrinsic GB blocking effect [40]. Moreover, the BSE-SEM images revealed that even the best SmFeAsO$_{0.85}$ bulk had non-superconducting Fe-As and RE_2O_3 occupying at least three quarters of the REFeAsO$_{0.85}$ GBs, making the active current path certainly much smaller than the geometrical cross-section of the sample [40]. Further to reveal the active local current paths, combined low temperature laser scanning microscopy (LTLSM) and SEM studies had been made [44]. With the *SEM* images they are able to show significant micro-structural differences between various regions of the sample. It is revealed that insulating Sm$_2$O$_3$ has a small surface to volume ratio and is mostly located within SmFeAsO$_{0.85}$ grains, so it has the smallest effect on current transport. On the other hand the dark gray Fe-As phase wets many GBs, thus interrupting grain to grain supercurrent paths, which are further degraded by extensive cracking, sometimes at GBs.

Fig. 10. (a), (b) SEM images of pure MgB$_2$ and 10 wt%n-SiC added samples. Reprinted with permission from [Nanotechnology 19 (2008) 125708]

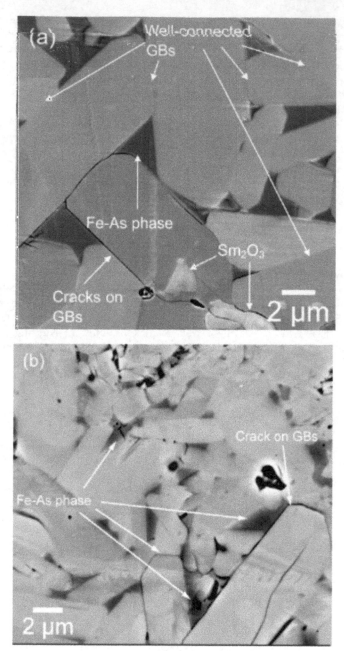

Fig. 11. BSE image of the (a) Sm and (b) Nd sample at high magnification. Although some grain boundaries are well connected, others are clearly obstructed by the Fe-As phase (dark contrast), Sm_2O_3 or Nd_2O_3 (white contrast) and cracks.

Reprinted with permission from [Supercond. Sci. Technol. 22 (2009) 015010].

Also, magnetic contamination is detected through *SEM* in the case of NdFeAsO single crystals grown out of NaAs flux under ambient pressure [45]. It is observed that some crystals show a lambda anomaly in the specific heat curve at ~12 K while the same is absent in others. They examined the cleaved (001) surfaces with *SEM* which were showing lambda anomaly. They turned to look at the edges of the crystals carefully. A 10-μm-thick layer was observed on the edges of the NdFeAsO crystals. With *EDS* it was found that some particles of TaAs, were surrounded by some very fine particles. After removal of that impurity the lambda anomaly disappeared. Thus with the help of *SEM* and *EDS* we can find out the actual cause which leads particular nature of a material.

B. Ba/Sr/K/Fe$_2$As$_2$ (122)

Soon after the discovery of 1111 family, other superconducting families based on FeAs layers (122, 111 and 11 structure) were reported, such as (Ba,K)Fe$_2$As$_2$ [46], LiFeAs [47], and FeSe [48]. Among all of these iron-based superconductors, the 122-type superconductors with a Tc of 38 K have a lower synthesis temperature and are oxygen free in comparison to the 1111-type. In addition, its Tc is much higher than those of the 111 and 11-type superconductors. SEM is used ingeniously to study the single crystal of Ba(Fe/Co)$_2$As$_2$ [16]. The topographic images of cleaved surface of optimally doped BaFe$_{1.8}$Co$_{0.2}$As$_2$ with *uniform* contrast have been observed in the secondary electron emission images acquired by SEM. They used secondary electron emission mode as it can register the contrast according to topography, chemical composition, and surface barrier (work function or ionization energy) of the sample [49] Small darker regions in the SEM image have been identified as marked by a rectangular box in Fig 12 (a). An SEM zoom-in image in the dark region reveals microscopic *domain* structures [Fig. 12]. They made resolved electrical transport measurements with use of SEM which have provided direct evidence of the coupling between superconductivity and local environment that is reflected by Co-concentration variation. In the uniform regions, the superconducting transition occurs at T_C = 22.1 K for 10% fixed percentage of the normal-state resistance. In the domain regions, although the onset superconducting transition temperature is found very close to that of the uniform regions, T_C varied over a broader range of 0.3-3.2 K. In addition, resistance of the domain regions above the transition onset temperature was noticed higher than that of the uniform region, indicating higher defect density in the domain regions.

Like HTSc's improvement in J_c is observed with increase of grain size in 122 systems also [50]. Effect of sintering temperature on the microstructure and superconducting properties of Sr$_{0.6}$K0.4Fe$_2$As$_2$ bulk samples was made. It was found that the annealing temperature had little influence on the critical temperature Tc. However, the irreversibility field H_{irr} and J_c were significantly affected by the sintering temperature. The *SEM* images reveal although samples had similar microstructure, the grain size increases monotonically as the sintering temperature rises. The grain size was less influenced by temperatures over 850 C. It was concluded the J_c enhancement may result mainly from better grain connectivity due to the decrease of impurity phases.

4. Limitations

Although SEM is very useful in finding grain size, their connectivity and then revealing various microscopic properties with physics behind that but there are some limiting conditions for it. It is not a complete characterization in the sense that it needs extra characterization techniques such as TEM, PL, XPS and MO to support the results. The sensitivity of SEM is known to be relatively poor for lighter elements such as B, C and O. Thus through EDS results, the actual percentage ratio cannot be determined as very light elements boron and carbon are lesser sensitive in comparison to others elements [15]. In nano-TiO$_2$ doped samples [51] almost similar micrographs for all samples was found irrespective of whether they are doped with n-TiO$_2$ or not. In their study with HRTEM it was concluded that several black holes that appear in the image are presumably the n-TiO$_2$.

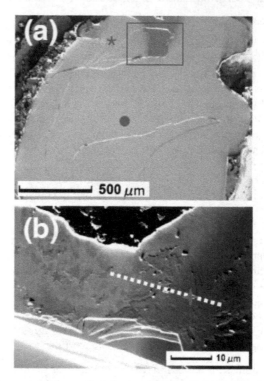

Fig. 12. Topographic images of cleaved surfaces of BaFe$_{1.8}$Co$_{0.2}$As$_2$ single crystal. (a) SEM image showing uniform contrast with some dark regions near the edge of crystal, as marked by a rectangular box. (b) Zoom-in SEM image showing domain structures. Marked regions by symbols and a dash line in (a) and (b) indicate where the transport measurements and composition probing are carried out.
Reprinted (Fig.) with permission from [T.-H. Kim, R. Jin, L. R. Walker, J. Y. Howe, M. H. Pan, J. F. Wendelken, J. R. Thompson, A. S. Sefat, M. A.McGuire, B. C. Sales, D. Mandrus, A. P. Li, Phys. Rev. B 80 (2009) 214518].

5. Summary

Summarily we can see that *SEM* has been widely used to explore the superconducting behavior. Grain size matters a lot in deciding the superconducting parameters of cuprate

HTSc. Some of the important aspects with which *SEM* deals with, is the grain size, morphology and alignment, structural defects, chemical composition. It has been used widely to explore from HTSc's, diborides to pnictides. With the time ways to use *SEM* and to extract information from superconductors got improved. Earlier for HTSc's it was used simply in finding grain size, grain connectivity and to figure out impurity regions. With these parameters superconducting behaviour was explained. Along with this in diborides it was also used to figure out the pinning centers and hence to enhance the applicable parameters. In Pnictides with spatially resolved electrical transport measurements it provided direct evidence of the coupling between superconductivity and local environment variation. We suppose in future SEM will be more plausible to understand so that superconductivity is better understood and improved.

6. Acknowledgements

Author S. K. Singh would like to acknowledge *CSIR*, India for providing fellowships. We are very much thankful NPG [Macmillan Publishers Ltd: NATURE 410 (2001) 186], IOP Publishing Ltd.{(SUST [Supercond. Sci. Technol. 22 (2009) 015010, Supercond. Sci. Technol. 22 (2009) 034012] and Nanotechnology **19** (2008) 125708)}, and APS [Phys. Rev. B 80 (2009) 214518] for providing permissions for the reprints of the images. We want to acknowledge the authors of the Ref. No. [15], [16], [20], [35], [40] for giving their consent to re-use the images.

7. References

[1] J. W Ekin, Adv. Ceram. Matter. 2 (1987) 586
[2] A Rosko, Y. M Chiang, J. S Moodera and D. A Rudman, American Ceram. Socie. (1988) 308
[3] D. S smith, S. Suasmoro, C.Gault, F. Caillaud and A. Smith, Revue Phys. Appl. 25 (1990) 61
[4] http://www.astm.org/
[5] http://www.astm.org/Standards/E112.htm
[6] T. Nagano, Y. Tomioka, Y. Nakayama, K. Kishio, K. Kitazawa, Phys. Rev. B 48 (1993) 9689
[7] S X Dou *et al.* Appl. Phys. Lett. 81 (2002) 3419
[8] W. K. Yeoh *et al.* Supercond. Sci. Technol. 19 (2006) 596
[9] H. Yamada, *et al.* Supercond. Sci. Technol. 19 (2006) 175
[10] C H Cheng, *et al.* Supercond. Sci. Technol. 16 (2003) 1182
[11] Senkowicz *et al.* Appl. Phys. Lett. 86 (2005) 202502
[12] R H T Wilke, *et al.* Phys. Rev. Lett. 92 (2004) 217003
[13] Xiang *et al.* Physica C 386 (2003) 611
[14] A. Matsumoto, *et al.* Supercond. Sci. Technol. 16 (2003) 926
[15] Arpita Vajpayee, V P S Awana, G L Bhalla and H Kishan Nanotechnology 19 (2008) 125708
[16] T.-H. Kim, R. Jin, L. R. Walker, J. Y. Howe, M. H. Pan, J. F. Wendelken, J. R. Thompson, A. S. Sefat, M. A. McGuire, B. C. Sales, D. Mandrus, A. P. Li, Phys. Rev. B 80 (2009) 214518
[17] Charles P. Poole, Jr., Handbook of superconductivity, Academic Press, California (2000)
[18] D. Sharma, Ranjan Kumar, H. Kishan and V.P.S. Awana, J Supercond Nov Magn 24 (2011) 205
[19] J. Kumar *et al.* J Supercond Nov Magn 23(2010) 493
[20] A Sotelo *et al.* Supercond. Sci. Technol. 22 (2009) 034012

[21] J.C. Bednorz and K.A. Müller, Zeitschrift für Physik B. 64(2) (1986) 189
[22] Chiang, Y.-M., Rudman, D.A., Leung, D.K., Ikeda, J.A.S., Roshko, A., Fabes, B.D., Physica C 152 (1988) 77
[23] J.W. Ekin *et al.* J. Appl. Phys. 62, (1987) 4821
[24] Shiming Zhou, Jiyin Zhao, Songnan Chu, Lei Shi Physica C 451 (2007) 38.
[25] R.J. Cava, Science 247 (1990) 656
[26] R. Gagnon, C. Lupien, and L. Taillefer, Phys. Rev. B 50 (1994) 3458
[27] M. Karppinen, V.P.S. Awana, Y. Morita, H. Yamauchi, Physica B, 312 (2003) 62
[28] P.R. Slater, C. Greaves: Physica C 180 (1991) 299
[29] Shiva Kumar, Anjana Dogra, M. Husain, H. Kishan and V.P.S. Awana, J. Alloys and compd. 352 (2010) 493
[30] N.P Liyanawaduge, Shiva Kumar Singh, Anuj kumar, V.P.S Awana and H.Kishan, J Sup. and Novel Magn doi: 10.1007/s10948-010-1063-7
[31] X.S. Wu, W.S. Tan, Y.M. Xu, E.M. Zhang,J. Du, A. Hu, S.S. Jiang, J. Gao Physica C 398 (2003) 131
[32] J. M. Tarascon *et al.* Phys. Rev. B 38 (1988) 8885
[33] O. Monnereau, Z.C. Kang, E. Russ, I. Suliga, G. Vacquier, T. Badéche, C. Boulesteix, A. Casalot Applied Superconductivity April (1995) 197
[34] Yeshurun, Y., Malozemoff, A. P. & Shaulov, A. Magnetic relaxation in high-temperature superconductors. Rev. Mod. Phys. 68 (1996) 911
[35] L. D. Larbalestier *et al.* NATURE 410 (2001) 186
[36] Manashi Nath and B. A. Parkinson Adv. Mater. 18 (2006) 1865
[37] S X Dou *et al.* Phys. Rev. Lett. 98 (2007) 097002
[38] Y. Kamihara, T. Watanabe, M. Hirano, H. Hosono, J. Am. Chem. Soc. 130 (2008) 3296
[39] S. Kaciulis *et al.* Surf. Interface Anal. 42 (2010) 692.
[40] F. Kametani, A. A. Polyanskii, A. Yamamoto, J. Jiang, E. E. Hellstrom, A. Gurevich, D. C. Larbalestier, Z. A. Ren, J. Yang, X. L. Dong, W. Lu, and Z. X. Zhao, Supercond. Sci. Technol. 22 (2009) 015010.
[41] A. Yamamoto, J. Jiang, C. Tarantini, N. Craig, A. A. Polyanskii, F. Kametani,F. Hunte, J. Jaroszynski, E. E. Hellstrom, D. C. Larbalestier, R.Jin, A. S. Sefat, M. A. McGuire, B. C. Sales, D. K. Christen, and D. Mandrus, Appl. Phys. Lett. 92 (2008) 252501
[42] B. Senatore, G. Wu, R. H. Liu, X. H. Chen, and R. Flukiger, Phys. Rev. B 78 (2008) 054514
[43] R. Prozorov, M. E. Tillman, E. D. Mun, and P. C. Canfield, New J. Phys.11 (2009) 035004
[44] F. Kametani, P. Li,1 D. Abraimov, A. A. Polyanskii, A. Yamamoto, J. Jiang, E. E. Hellstrom, A. Gurevich, D. C. Larbalestier, Z. A. Ren, J. Yang, X. L. Dong, W. Lu, and Z. X. Zhao Appl. Phys. Lett. 95 (2009) 142502
[45] J.-Q. Yan, Q. Xing, B. Jensen, H. Xu, K. W. Dennis, R. W. McCallum, and T. A. Lograsso Phys. Rev. B. 84 1250
[46] M. Rotter, *et al.* Phys. Rev. B 78 (2008) 020503
[47] X C Wang, *et al.* Solid State. Commun. 148 (2008) 538
[48] Hsu F C *et al.* Proc. Natl Acad. Sci. USA 105 (2008) 14262
[49] Ma Y W, Gao Z S, Wang L, Qi Y P, Wang D L and Zhang X P Chin. Phys. Lett. 26 (2009) 037401
[50] Zhiyu Zhang, Yanpeng Qi, Lei Wang, Zhaoshun Gao, DongliangWang, Xianping Zhang and Yanwei Ma Supercond. Sci. Technol. 23 (2010) 065009
[51] H. Kishan, V.P.S. Awana, T.M. de Oliveira, Sher Alam, M. Saito, O.F. de Lima Physica C 458 (2007) 1

13

Synthesis and Characterisation of Silica/Polyamide-Imide Composite Film for Enamel Wire

Xiaokun Ma and Sun-Jae Kim[*]

*Institute/Faculty of Nanotechnology and Adv. Materials Engin.,
Sejong University #98 Gunja-dong, Gwangjin-gu, Seoul,
South Korea*

1. Introduction

In the past decade, the demand for polyamide-imide (PAI) and other high-temperature resistant polymeric materials has grown steadily because of their outstanding mechanical properties and excellent thermal and oxidative stability (Zhong, 2002; Sun, 2006; Yanagishita, 2001; Babooram, 2008). PAI is well-known for its low thermal expansion coefficient and dielectric constant. In microelectronics, PAI has been widely used as an inter-dielectric material, and in the large-scale integrated circuit industry, as an electrical insulation for conventional appliances (Kawakami, 1996, 1998, 2003; Rupnowski, 2006; Wu, 2005). Compared with pure polyimide and polyamide, PAI exhibits better process ability and heat-resistant properties. The application of PAI as a wire-coating material with thermal-resistant properties has attracted increasing interest (Chen, 1997; Ranade, 2002; Ma, 2007). However, with the introduction of higher-surge voltage devices, an increasing number of insulation electric breakdown cases have been reported. Insulation electric breakdown must be prevented because it may lead to electrical component failure or may endanger the people handling the component. Thus, the development of an organic/inorganic composite insulating material is essential in designing insulation for continuous use (Alexandre, 2000; Hossein, 2007; David, 1995; Yang, 2006). Polymer composites have received much attention, as various properties of the original matrix polymer can be considerably improved by adding a limited percentage of inorganic filler (Jiao, 1989; Rangsunvigit, 2008; Xu, 2007; Hwang, 2008; Kim, 2007; Rankin, 1998).

Silica has been commonly used as an inorganic component because it is effective in enhancing the mechanical and thermal properties of polymers. Various studies on the preparation of polymer/silica composite films have been conducted (Butterworth, 1995; Mosher, 2006; Kim, 2006; Ahn, 2006; Stathatos, 2004).

The properties of hybrid composites are affected by many factors, such as particle size, size distribution, and filler content. In addition, the inorganic particle shape, surface structure,

---

[*] Corresponding Author

and mechanical properties of a filler (stiffness and strength, among others) play important roles in inorganic/organic composite material synthesis. In particular, the bond strength between the inorganic particles and the polymer matrix, influenced by the dispersion aid type or coupling agent used, should be improved (Kusakabe, 1996; Fuchigami, 2008; Castellano, 2005; Alexandre, 2000; Wu, 2006; Zheng, 2007; Ohki, 2005).

Silica nanoparticles, as important inorganic materials, have emerged as an area of intense interest because of their special physical and chemical properties, such as their small size, strong surface energy, high scattered performance, and thermal resistance (Ouabbas, 2009; Lee, 2006; Bhagat, 2008; Oh, 2009; Xue, 2009). However, the applications of silica nanoparticles are largely limited because of their highly energetic hydrophilic surface, which causes the silica nanoparticles to easily agglomerate. Surface modification methods using different surfactant agents may resolve this limitation. Thus, the strong interface adhesion between the organic matrix and the silica nanoparticles is the key to the application of silica nanoparticles as fillers.

Jadav et al. successfully synthesised a silica/polyamide nanocomposite film via interfacial polymerisation using two types of silica nanoparticles of 16 and 3 nm in size (Jadav, 2009). The nanocomposite films exhibited superior thermal stability to the pure polyamide membranes. In the current work, silica nanoparticle loading significantly modified the polyamide network structure, pore structure, and transport properties. The excellent membrane performance in terms of separation efficiency and productivity flux was also discussed. Zhang et al. prepared a novel isometric polyimide/silica hybrid material via sol-gel technique (Zhang, 2007). Initially, 3-[(4-phenylethynyl) phthalimide] propyl triethoxysilane was synthesised to modify the nanosilica precursor. Then, the isomeric polyimide/silica hybrid material was produced using isomeric polyimide resin solution and the modified nanosilica precursor after heat treatment. The isomeric polyimide/silica composite has much better thermal properties and nano-indenter properties than those of the isomeric polyimide.

In the current work, the commercial silica nanoparticles and self-synthesised spherical silica particles were successfully dispersed in the PAI polymer matrix after the surface modification process. The cationic surfactant cetyltrimethyl ammonium bromide (CTAB) was chosen to modify the silica nanoparticles. The amount of CTAB added in modifying the silica nanoparticles was increased from 0 to 3 wt%. After the surface modification process, the CTAB-modified silica nanoparticles showed better compatibility with the PAI polymer matrix. The results indicate that CTAB plays an important role in the preparation of silica/PAI composite film. The thermal stability improved and the decomposition temperature increased with increasing amounts of silica particles. The thermal expansion coefficient of the composite film was lower than that of the PAI polymer matrix, which is helpful in extending the life of the enameled wire.

2. Synthesis and characterisation of the spherical silica/PAI composite film

Spherical silica/PAI composite films have been successfully prepared via simple ultrasonic blending. In the current study, the spherical silica particles were prepared according to the Stöber procedure, and the size was controlled to approximately 300 nm at room temperature. After the surface modification process, the commercial silica nanoparticles and the self-synthesised spherical silica particles were dispersed in separate PAI polymer matrices. The correlation of the silica particle size with the amount of silica dispersed in the PAI under optimal experimental conditions was discussed.

2.1 Preparation of the silica/PAI composite film

2.1.1 Synthesis of the spherical silica submicron particles

The spherical silica particles were prepared according to the Stöber procedure, which allows the preparation of monodispersed silica particles with particle sizes in the nanometer to submicron range. A 500 ml three-necked flask equipped with a mechanical stirrer was filled with 45 ml ethanol, 2 ml NH_3 H_2O, and 1 ml deionised H_2O. The spherical silica synthesis was initiated by the rapid addition of 2 ml tetra-ethoxy-silane (TEOS) to a stirred solution. After the mixture was vigorously stirred for 2 h, spherical silica submicron particles of approximately 300 nm in size, were obtained at room temperature. Then, the silica particles were centrifuged at 10000 rpm for 30 min. The resultant silica particles were washed with ethanol and distilled H_2O, and then modified with CTAB.

2.1.2 Surface modification of the silica particles

CTAB can be directly added into the submicron silica solutions, and the CTAB amount can be increased from 0 to 3wt % at the optimal temperature of 60 °C. After the surface modification process, the modified silica particles were collected via centrifugation and then dried at 60 °C.

In the experiment, 100 ml deionised H_2O and 1 g commercial silica nanoparticles were added to a flask and the solution was adjusted to pH 8 by the addition of 0.1 M NaOH. The silica nanoparticles were modified at 65 °C with the addition of CTAB under constant stirring. The dispersal state of the silica nanoparticles in the PAI matrix was improved by increasing the amount of CTAB from 0 to 3.0 wt%. After the surface modification process, the modified nanosilica particles were collected via suction filtration and then dried at 90 °C for 6 h.

2.1.3 Preparation of the silica/PAI composite film

The nanosilica/PAI composite films were prepared via simple ultrasonic blending. Two grams of PAI powder were dissolved in 3 ml N,N-dimethyl form amide (DMF). The silica nanoparticles were added into the solution, and the amount of silica was increased from 2 to 10 wt%. The mixture was put under ultrasonic dispersion for approximately 3 h at room temperature. The mixture solution was then cast on a square glass plate (5 cm × 5 cm) using a bar coater. Bar coaters are primarily used in applying a variety of coatings or emulsions to a multitude of substrates. The bar coater used in the current experiment had a 0.2 mm diameter and was made of stainless steel wire, which resulted in more uniform silica/PAI composite films. The films were initially heated to remove the solvent in the vacuum oven. The temperature was controlled as follows: at 60 °C for 4 h, at 80 °C for 2 h, increased to 120 °C for 1 h, and finally, kept at 160 °C for 1 h. The experimental details of the silica/PAI composite film preparation are shown in Scheme 1.

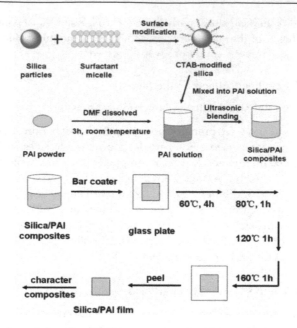

Scheme 1. Preparation of the silica/PAI nanocomposite films.

The spherical silica/PAI composite films were obtained under the same method. Spherical submicron silica particles do not easily agglomerate, so the amount of submicron silica added to PAI in the system was increased to 25%. The reactants of the submicron silica/PAI composites are listed in Table 1.

Reactant Sample	Silica Particles (g)	PAI (g)	DMF (ml)	Theoretical Weight Percent (wt%)
Composite 1	0.04	2.0	3	1.96
Composite 2	0.08	2.0	3	3.85
Composite 3	0.12	2.0	3	5.66
Composite 4	0.16	2.0	3.5	7.41
Composite 5	0.20	2.0	3.5	9.10
Composite 6	0.30	2.0	3.5	13.0
Composite 7	0.40	2.0	4	16.7
Composite 8	0.50	2.0	4	20.0

Table 1. Reactants of the different submicron silica/PAI composite samples.

2.2 Characterisation of the silica/PAI composite film

The fracture surfaces of the composite films were studied using a scanning electron microscope (SEM Hitachi S-4700, Hitachi Co.). Prior to SEM imaging, the samples were sputtered with thin layers of Pt-Pd. The silica/PAI nanocomposite films were characterised by an FT-IR (OMNIC NICOLET 380) spectrometer. The spectra were measured in the range 4000–650 cm^{-1}. A Scinco STA S-1500 simultaneous thermal analyser was then used to analyse the thermal stability of the nanosilica/PAI composite films. The samples were heated from 30 to 800 °C at 10 °C/min under air atmosphere. The coefficients of thermal expansion (CTE) of the silica/PAI composites films were evaluated using a Q 400 EM (U.S.A) thermomechanical analyser (5 °C/min from 25 to 300 °C, 50 mN). All the samples were 3 mm × 16 mm, cut from the original films using a razor blade.

2.2.1 CTAB effect on the synthesis of silica/PAI composite film

The FT-IR spectra of the silica/PAI composite films are shown in Fig. 1. The effect of the surfactant on the composite films was evaluated by increasing the CTAB dosage from 0 to 3 wt%. The amount of silica nanoparticles added to the PAI was 6 wt%. The character vibrations of the Si-O were observed at 1086, 945, and 796 cm^{-1}, as shown in Fig. 1 (e). After the surface modification process, the typical stretching vibrations of the C-H were found at 2855 and 2928 cm^{-1}, which resulted from the –CH$_2$ and –CH$_3$ in the CTAB. Figure 1 shows the typical characteristic bands of the PAI polymer matrix that were found, such as the N-H stretching band at 3317 cm^{-1}, the amide C=O region at around 1710 cm^{-1}, and the bands at 1771 and 1710 cm^{-1} associated with the imide carbonyl band. The bands are similar to one another because the same amount of silica was added into the composites. The characteristic stretching vibration of Si-O at 1086 cm^{-1} became wider when the silica nanoparticles were modified by CTAB. This peak broadening may be explained by the organic side-chain of CTAB grafted on the surface of the silica nanoparticles, which improved the interaction between the silica nanoparticles and the PAI polymer matrix.

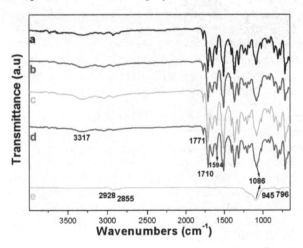

Fig. 1. FT-IR spectra of (a) unmodified-silica/PAI nanocomposites, (b) 1% CTAB-silica/PAI nanocomposites, (c) 2% CTAB-silica/PAI nanocomposites, (d) 3% CTAB-silica/PAI nanocomposites, and (e) CTAB-modified silica nanoparticles.

The fracture surface micrographs of the silica/PAI nanocomposite films are shown in Fig. 2. In Fig. 2 (a), some silica agglomerations are found in the micrograph when the unmodified-silica nanoparticles were added into the PAI matrix. The silica nanoparticles easily agglomerate because of their large surface-to-volume ratios and high surface tension. However, the CTAB-modified silica nanoparticles dispersed well in the PAI matrix with the increase in CTAB. The dispersal state of the silica nanoparticles improved when the silica nanoparticles were modified with 1 wt% CTAB, as shown in Fig. 2 (b). In Fig. 2 (c), the silica nanoparticles are almost monodispersed, and little agglomeration is observed when 2 wt% CTAB-modified silica nanoparticles were added into the PAI polymer. After the silica were modified with 3 wt% CTAB, the silica nanoparticles became monodispersed without any agglomerations, although the amount of silica nanoparticles added to the PAI was increased to 6 wt%, as shown in Fig. 2 (d). CTAB improves the dispersal state of the silica nanoparticles in a PAI polymer matrix, with an optimal dosage of 3 wt%.

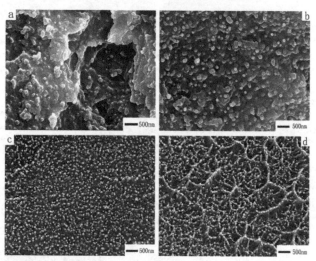

Fig. 2. Fracture surface micrographs of (a) unmodified-silica/PAI nanocomposites, (b) 1% CTAB-silica/PAI nanocomposites, (c) 2% CTAB-silica/PAI nanocomposites, and (d) 3% CTAB-silica/PAI nanocomposites.

Fig. 3 shows the thermogravimetric analysis (TGA) plots of the different CTAB modified-silica/PAI nanocomposites films when 6 wt% nanosilica was added into the PAI matrix. Some differences are found in the weight loss curves shown in Fig. 3. When the temperature was increased to 475 °C, the silica/PAI composites began to decompose. Compared with the unmodified-silica/PAI composites, the decomposition temperature increased after the silica nanoparticles were modified by CTAB. The results indicate that the CTAB-modified silica particles improve the thermal stability of the PAI polymer matrix. In addition, the decomposition temperature of composite films increased with increasing CTAB dosage. When the amount of CTAB added to the silica was increased from 0 to 3 wt%, the decomposition temperature of the composite films increased from 646 to 658, 671, and 682 °C, respectively. A thermal decomposition process occurs when the temperature approaches 595 °C, as shown in Figs. 3 (b)-(d). Therefore, the interaction between the silica nanoparticles and PAI polymer matrix is enhanced after the silica are modified by CTAB.

CTAB improves not only the dispersal state of silica in the PAI matrix, but also the thermal stability of the composite film because of the better interaction between the nanosilica and

PAI matrix. The amount of silica in the composites was calculated based on the plots, and the result was about 5.6 wt%, which is in accordance with the theoretical data.

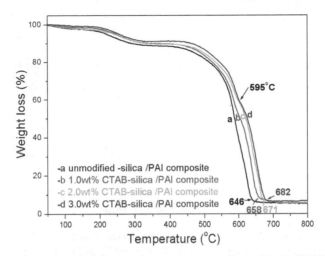

Fig. 3. TGA plots of (a) unmodified-silica/PAI nanocomposites, (b) 1% CTAB-silica/PAI nanocomposites, (c) 2% CTAB-silica/PAI nanocomposites, and (d) 3% CTAB-silica/PAI nanocomposites.

Thus, CTAB, with an optimal dosage of 3 wt%, was chosen to modify the nanosilica. The fracture surface micrographs of the composites show that the silica nanoparticles were well-dispersed in the PAI matrix after the surface modification process. In the TGA plots of the silica/PAI nanocomposites, the thermal stability and the decomposition temperature increased with increasing CTAB. Therefore, CTAB is important in the preparation of the silica/PAI nanocomposite film.

2.2.2 The effect of silica nanoparticle amount on the properties of silica/PAI composite film

The FT-IR spectra of the pure PAI and some silica/PAI nanocomposite films are shown in Fig. 4 The amount of silica nanoparticles added to PAI was changed from 2 to 10 wt%, and the dosage of CTAB added to the silica was 3 wt%. In the spectra shown in Fig. 4, the bands at 1771 and 1710 cm^{-1} are associated with the imide carbonyl band. Both bands are insensitive to the presence of the silica nanoparticles. The bands in the region from 945 to 650 cm^{-1} increased with the increase in silica content, caused by the presence of a broad band associated with the vibration of the Si-O bond. The N-H stretching band at 3317 cm^{-1} was slightly intensified with the increase in silica content, indicating the hydrogen-bonded N-H groups in the PAI polymer and the Si-O-Si or Si-O-H groups of the silica nanoparticles. The characteristic band at 1594 cm^{-1} comes from the benzene-ring stretch and a contribution from the O-H bond in monomeric H_2O, which also has a band at 1663 cm^{-1}. All the characteristic peaks in the composites indicate that the interaction between the silica nanoparticles and PAI polymer matrix is sensitive to the amount of silica in the composites.

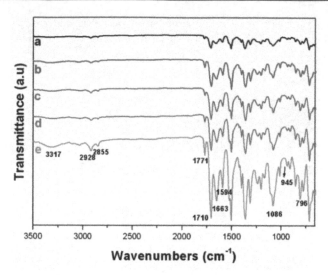

Fig. 4. FT-IR spectra of the silica/PAI composites with the amount of silica nanoparticles added to PAI at (a) 2 wt%, (b) 4 wt%, (c) 6 wt%, (d) 8 wt%, and (e) 10 wt%.

The fractured surface micrographs of the pure PAI and several composite films are shown in Fig. 5. The fracture surfaces of the pure PAI film are uniform, and the continuous polymer phase is shown in Fig. 5 (a). Figs. 5 (b)–(f) show the fracture surfaces of the different silica/PAI composites. The amount of silica added to the PAI was increased from 2 to 10 wt%. The larger the amount of silica nanoparticles added to the PAI, the greater their amount found in the fracture surface micrographs. In Fig. 5 (f), when 10 wt% silica nanoparticles was added into the PAI, the silica nanoparticles remained monodispersed without any agglomerations. The results indicate that the CTAB-modified silica nanoparticles have better dispersal state in the PAI polymer matrix. In addition, the CTAB-modified silica nanoparticles increased the silica nanoparticle content in the composites. Thus, the surface modification process is an effective method of preparing silica/PAI nanocomposites.

The thermal stability of the silica/PAI composite films was evaluated via TGA. The TGA plots of the PAI and the composites with the different amounts of silica nanoparticles are shown in Fig. 6. The amount of silica nanoparticles added to the PAI polymer matrix was increased from 0 to 10 wt%. The plots are shown in Figs. 6 (a)–(f). A weight loss is observed above 170 °C on all the TGA plots, which corresponds to water and solvent losses. When the temperature was increased to 450 °C, the PAI matrix began to decompose. The decomposition temperature increased when the silica nanoparticles were added into the PAI polymer matrix. At the same temperature, all the curves of the composites indicated that the composite weight loss was less than that of the pure PAI matrix. The silica/PAI composites have higher decomposition temperature when the PAI polymer matrix loses the same weight. The thermal stability of PAI was enhanced with the increase in the silica content. The amount of silica nanoparticles in the composites were calculated accurately in the TGA plots. The silica content in the composites based on Figs. 6 (b)–(f) are 1.9, 3.6, 5.8, 7.3, and 10.2 wt%, respectively.

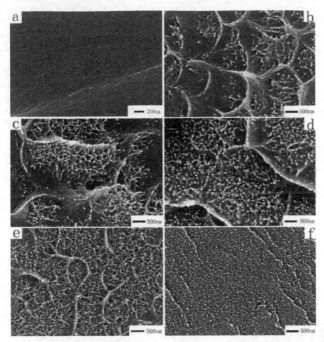

Fig. 5. Fracture surface micrographs of (a) pure PAI and silica/PAI nanocomposite, with the amount of silica added to PAI at (b) 2 wt%, (c) 4 wt%, (d) 6 wt%, (e) 8 wt%, and (f) 10 wt%.

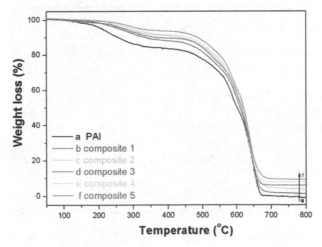

Fig. 6. TGA plots of the (a) PAI matrix, and the silica/PAI nanocomposite films with the amount of silica added to PAI at (b) 2 wt%, (c) 4 wt%, (d) 6 wt%, (e) 8 wt%, and (f) 10 wt%.

CTE is an important parameter in evaluating the properties of enamel wire. A low CTE can reduce thermal stress build-up and prevent device failure through peeling and cracking at the interface between the polymer film and the copper. The CTE curves of the pure PAI film and some composite films are shown in Fig. 7. The CTE value of the PAI films was 3.87 × 10⁻⁵ m/m/°C, whereas that of the silica/PAI composite film decreased to 3.69 × 10⁻⁵ and 3.51 × 10⁻⁵m/m/°C when the amount of silica added to PAI was 4 and 6 wt%, respectively. The CTE values continuously decreased with increasing amount of the silica particles. In particular, the CTE value decreased to 3.35 × 10⁻⁵ m/m/°C when the amount of silica added to PAI was 10 wt%, as shown in Fig. 7 (d). Compared with the PAI polymer film, the silica/PAI composite films had lower CTE, which may be attributed to the rigidity and stiffness of the silica nanoparticles and the interaction between the silica and PAI polymer matrix. The rigidity and stiffness of the silica nanoparticles limit the polymer chain movement, resulting in the decrease of the PAI matrix thermal expansion.

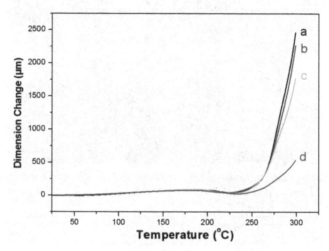

Fig. 7. CTE curves of the (a) pure PAI film and the composite films with silica content at (b) 4 wt%, (c) 6 wt%, and (d) 10 wt%.

The thermal stability of the silica/PAI nanocomposites improved and the decomposition temperature increased when the amount of silica nanoparticles was increased. The lower CTE of the composite films can reduce the peeling and cracking at the interface between the polymer film and the copper. In the current system, the high thermal stability and low CTE show that the silica/PAI nanocomposite films can be widely used in the enamel wire industry.

2.2.3 Effect of silica diameter on the properties of the silica/PAI composite film

In the past years, innovative inorganic/organic composite technology has been used to develop wires with better inverter-surge-resistance and mechanical properties than those of conventional enamelled wires. Kikuchi et al. (Kikuchi, 2002) emphasised that the inorganic/organic composite film can decrease erosion rate by increasing creeping distance and decreasing collision energy via reflection or scattering, as shown in Scheme 2.

Submicron spherical silica particles were added into the PAI polymer matrix during the synthesis of the silica/PAI composite films to evaluate the effects of silica diameter on their properties. The characterisation results of the spherical silica/PAI composites are discussed as follows.

Fig. 8 shows the SEM micrographs of (a) the spherical silica submicron particles and (b) the fracture surface of the pure PAI and several spherical silica/PAI composite films. In Fig. 8 (a), the mean diameter of the spherical silica nanoparticles is about 300 nm, and most are well-dispersed after the surface modification process. The fracture surfaces of the pure PAI film are uniform, and the continuous polymer phase is shown in Fig. 8 (b). Figs. 8 (c)–(f) show the fracture surfaces of the different silica/PAI composites. The amount of silica added to the PAI was increased from 2 to 8 wt%. When the submicron spherical silica was added into the PAI matrix, some prominent features were observed on the fracture surfaces of the composite films, as shown in Fig. 8 (c). The spherical silica particles were embedded in the PAI matrix, and the continuous PAI organic phase appeared when the amount of silica added to the PAI was 2 and 4 wt%, as shown in Figs. 8 (c) and (d). When the amount of submicron silica increased, more spherical silica particles were observed on the fracture surfaces of the composite films. The continuous organic PAI phase separated, as shown in Fig. 8 (e). Some alveolate pores were observed in Fig. 8(f), when 8 wt% spherical silica was added into the PAI matrix. These pores are caused by the removal of the submicron silica particles from the PAI matrix when the composite films were broken.

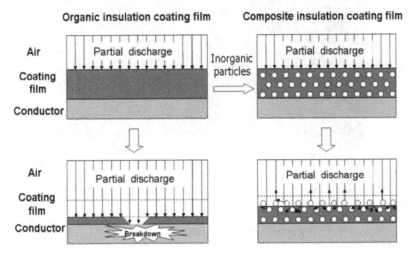

Scheme 2. Mechanism of erosion suppression.

More submicron spherical silica particles were subsequently added into the PAI matrix. The SEM micrographs of the composite film fracture surfaces are shown in Fig. 9. The micrographs show that the diameters of the submicron silica particles are uniform. In addition, the spherical silica particles are orderly arranged in the PAI matrix. An increase in the silica particles added into the films may result in a more compact framework. Partial discharge is a prime factor causing enamel wire breakdown, so the composite films caused a decrease in the erosion rate via reflection and scattering when the spherical silica particles were added into the PAI matrix. That is, the charged particles were reflected and scattered around the submicron silica, which slowed down the corrosion process. Therefore, the particle discharge resistance is improved with the increase in silica content. When the amount of silica added to the PAI was increased to 25 wt%, as shown in Fig. 9 (d), the fracture surfaces of the composite films remained well-integrated without the obvious phase separation. The results indicate that such a simple method effectively increases the amount of inorganic silica particles in the PAI matrix after the surface modification process.

Fig. 8. SEM micrographs of (a) spherical silica submicron particles and fracture surfaces of (b) pure PAI film, (c) composite 1 with 2 wt% submicron silica, (d) composite 2 with 4 wt% submicron silica, (e) composite 3 with 6 wt% submicron silica, and (f) composite 4 with 8 wt% submicron silica.

The FT-IR spectra of CTAB-modified silica, pure PAI, and some submicron silica/PAI composite films are shown in Fig. 10. As shown in Fig. 3 (a), typical bands of silica were observed at 1086, 950, and 809 cm^{-1}, indicating the stretching vibrations of Si-O. The stretching vibration peaks of C-H were found at 2861 and 2915cm^{-1}, which came from the -CH$_2$ and -CH$_3$ in CTAB. The FT-IR spectra of PAI show the presence of the imide carbonyl band at 1776 and 1715 cm^{-1}, and the peaks at 3320 and 1715 cm^{-1} come from the N-H stretching and C=O region, respectively. Other characteristic bands include the absorption at 1600 cm^{-1} caused by the benzene ring stretch and a contribution from the O-H bond in monomeric H$_2$O, which also has a band at 1663 cm^{-1}. The bands at 950–650 and 3320 cm^{-1} increased with the increase in silica content. Furthermore, the adsorption peak at 1086 cm^{-1} was intensified in the spectra of the composites. All the characteristic peaks in the composites were insensitive to the presence of the silica component, indicating good interaction between the spherical silica and the PAI polymer matrix.

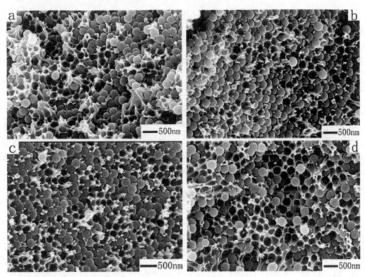

Fig. 9. SEM micrographs of (a) composite 5 with 10 wt% submicron silica, (b) composite 6 with 15 wt% submicron silica, (c) composite 7 with 20 wt% submicron silica, and (d) composite 8 with 25 wt% submicron silica.

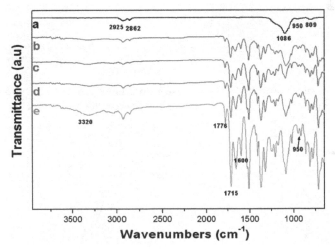

Fig. 10. FT-IR spectra of (a) CTAB-modified submicron silica particles, (b) pure PAI, (c) composite 1 with 2 wt% silica, (d) composite 2 with 4 wt% silica, and (e) composite 4 with 8 wt% silica.

The TGA plots of the PAI and the composites with the different amounts of submicron silica particles are shown in Fig. 11. Obvious weight loss is found in these plots. A weight loss is observed below 180 °C in all the TGA plots, corresponding to water and solvent losses. When the temperature was increased to 450 °C, the PAI matrix began to decompose with increasing decomposition temperature. The decomposition temperature increased from

640 to 678 °C after the silica particles were added to the PAI polymer matrix. The thermal stability of the PAI was increased by the incorporation of the submicron silica particles, as clearly shown in Fig. 11. The amount of silica particles in the composites were accurately calculated from the TGA plots. The silica content in the composites calculated based on the plots in Figs.11 (b)–(e) are 1.68, 3.80, 5.80, and 7.24 wt%. The calculated data are close to the theoretical values listed in Table 1.

Inorganic silica particles improve the thermal stabilities of the PAI polymer matrix and reduce the cost of enameled wire. To obtain the least expensive and most stable composite films, the amount of silica particles was increased from 10 to 25 wt%. In Fig. 12, the TGA curves are similar to the composite films shown in Fig. 11. The amount of silica particles was calculated based on the plots. The realistic data were 9.3, 13.7, 16.5, and 20.4 wt%.

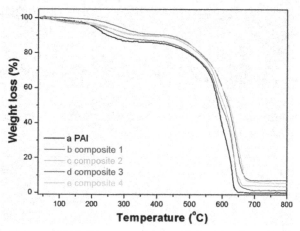

Fig. 11. TGA plots of (a) pure PAI, (b) composite 1 with 2 wt% silica, (c) composite 2 with 4 wt% silica, (d) composite 3 with 6 wt% silica, and (e) composite 4 with 8 wt% silica.

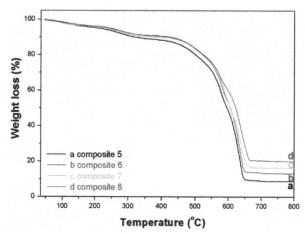

Fig. 12. TGA plots of (a) composite 5 with 10 wt% silica, (b) composite 6 with 15 wt% silica, (c) composite 7 with 20 wt% silica, and (d) composite 8 with 25 wt% silica.

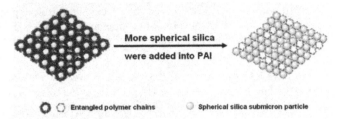

Scheme 3. Interactions between the silica submicron particles and the PAI polymer matrix.

Second, the thermal stability of the composite film was further improved by increasing the silica particles. However, the decomposition temperature was increased from 650 to 668 °C, the results of which did not exceed that of the composites shown in Fig. 11. These results indicate that the interaction between the silica and the PAI matrix changed when more silica submicron particles were added into the composite films. Initially, the thermal stability of PAI improved after the silica submicron particles were added into the PAI matrix. Given the thermal motion of molecules, the cohesion between the adjacent polymer chains and the partial resistance and friction from the incorporation of silica in the PAI matrix had to be overcome. When more silica submicron particles were added into the PAI polymer, the interaction between the silica and the PAI matrix weakened because of the more compact and orderly arrangement of silica in the PAI matrix, as shown in Scheme 3. In addition, the silica submicron particles were easier to break away from the fracture surfaces, as confirmed by the SEM micrographs shown in Fig. 9. As a result, the decomposition temperature of the composite film did not increase with increase in the amount of silica particles, when the amount of silica added to PAI was more than 10 wt%. However, the thermal stability and decomposition temperature of the composite film were obviously more prominent than those of the pure PAI polymer matrix. Considering the similar thermal stability and lower price of the enameled wire, the amount of silica submicron particles can be modulated from 4 to 25 wt%.

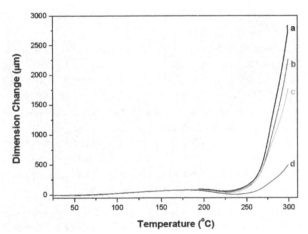

Fig. 13. CTE curves of the (a) pure PAI film, (b) composite 2 with 4 wt% silica, (c) composite 3 with 6 wt% silica, and (d) composite 6 with 15 wt% silica.

Controlling the CTE value of the dielectrical PAI is important because copper is the typical choice for defining the circuit lines. The selection of composites with the CTE close to that of copper is very critical, and can prevent damage to the copper interconnection; thus improving the reliability of the integrated circuits.

The CTE curves of the pure PAI film and some composite films are shown in Fig. 13. The CTE value of the pure PAI film is $3.87 \times 10^{-5} m/m/°C$, whereas that of the composite films in Figs. 13 (b)–(d) are 3.76×10^{-5}, 3.57×10^{-5}, and $3.25 \times 10^{-5} m/m/°C$, respectively. The CTE values continuously decreased with the increasing amount of the silica particles. The CTE value decreased by 16% when the amount of silica added to the PAI was 15 wt%. Such a variation in CTE is explained by the rigidity and stiffness of the silica submicron particles, which limit the movement of the polymer chain, resulting in the decrease of the thermal expansion of the PAI matrix. When the temperature increased to 220 °C, it approached the glass transition temperature of PAI. The dimension change significantly decreased with increased silica content. Therefore, the silica submicron particles effectively decrease the CTE value of the PAI polymer matrix, which in turn increases the thermal stress build-up, resulting in device failure through peeling and cracking at the interface between the PAI polymer film and the copper.

The thermal stability and the CTE value of the spherical silica/PAI composite films significantly improved when the submicron silica particles were added into the PAI polymer matrix. The submicron silica particles were obtained through the sol-gel method. Most previous studies used a sol-gel process because of the diameter of the silica particles. In the present study, the diameter effect was easily controlled using the Stöber procedure. Compared with the silica nanoparticles, the submicron silica particles have better dispersal state and do not easily agglomerate. Thus, more submicron silica particles are well-dispersed in the PAI polymer matrix after the surface modification process. Considering the higher thermal stability and lower CTE value, especially the lower cost, submicron silica/PAI composite films can be widely used in the enamel wire industry.

3. Conclusion

In the current work, silica particles with two different diameters were successfully added into the PAI polymer matrix in the synthesis of silica/PAI composite films via simple ultrasonic blending. First, the effect of CTAB on the synthesis of the silica/PAI composite film was investigated. The optimal dosage of CTAB is 3 wt%. The fracture surface micrographs of the composites show that the silica nanoparticles are well-dispersed in the PAI matrix after the surface modification process. In the TGA plots of the silica/PAI nanocomposites, the thermal stability and the decomposition temperature obviously increased with increasing CTAB dosage. Therefore, CTAB improves not only the dispersal state of silica in the PAI matrix, but also the thermal stability of the composite film because of better interaction between the nanosilica and the PAI matrix.

When the amount of silica nanoparticles added to the PAI was increased from 2 to 10 wt%, the thermal stability of the silica/PAI nanocomposites improved. The decomposition temperature increased with the increase in the amount of silica nanoparticles. In particular, the CTE value decreased when the silica particles were added into the PAI matrix. The CTE value further decreased with the increase in the amount of silica nanoparticles. When the submicron silica particles were added into the PAI polymer matrix, a similar conclusion was reached. However, more submicron silica particles are well-dispersed in the PAI polymer matrix after the surface modification process because the submicron silica particles have better dispersal state. Considering the higher thermal stability, lower CTE value, and the lower cost, silica/PAI composite films can be widely used in the enamel wire industry.

4. Acknowledgement

This research was supported by Basic Science Research Program through the National Research Foundation of Korea (NRF) Funded by the Ministry of Education, Science and Technology. (No. 2011-0016699). Also, it was supported by Sanhak Fellowship program funded of Korea Sanhak foundation.

5. References

Ahn, B. Y.; Seok, S. I.; Hong, S.; Oh, J.; Jung, H.; Chung, W. J. (2006). Optical properties of organic/inorganic nanocomposite sol-gel films containing LaPO$_4$:Er,Yb nanocrystals. *Optical Materials*, 28, 4, pp. 374-379

Alexandre, M.; Dubois, P. (2000). Polymer-layered silicate nanocomposites: preparation, properties and uses of a new class of materials. *Materials Science and Engineering*, 28, 1-2, pp. 1-63

Babooram, K.; Francis, B.; Bissessur, R.; Narain R. (2008). Synthesis and characterization of novel (amide–imide)-silica composites by the sol–gel process. *Composites Science and Technology*, 68, 3-4, pp. 617–624

Bhagat, S. D.; Kim, Y.; Suh, K.; Ahn, Y.; Yeo, J.; Han, J. (2008). Superhydrophobic silica aerogel powders with simultaneous surface modification, solvent exchange and sodium ion removal from hydrogels. *Microporous and Mesoporous Materials*, 112, 1-3, pp. 504-509

Butterworth, M. D.; Corradi, R.; Johal, J.; Lascelles, S. F.; Maeda, S.; Armes, S. P. (1995). Zeta Potential Measurements on Conducting Polymer-Inorganic Oxide Nanocomposite Particles. *Journal of Colloid and Interface Science*, 174, 2, pp. 510-517

Castellano, M.; Conzatti, L.; Costa, G.; Falqui, L.; Turturro, A.; Valenti, B.; Negroni, F. (2005). Surface modification of silica: 1. Thermodynamic aspects and effect on elastomer reinforcement. *Polymer*, 46, 3, pp. 695-703

Chen, L. W.; Ho, K. S. (1997). Synthesis of Polyamide-imide by Blocked-Methylene Diisocyanates. *Journal of Polymer Science Part A: Polymer Chemistry*, 35, 9, pp. 1711-1717

David, I. A.; Scherer, G. W. (1995). An Organic/Inorganic Single-Phase Composite. *Chemistry of Materials*, 7, pp. 1957-1967

Fuchigami, K.; Taguchi, Y.; Tanaka, M. (2008) Synthesis of spherical silica particles by sol-gel method and application. *Polymer for Advanced Technologies*, 19, pp. 977-983

Hossein, S. S.; Lia, Y.; Chunga, T.; Liu, Y. (2007). Enhanced gas separation performance of nanocomposite membranes using MgO nanoparticles. *Journal of Membrane Science,* 302, 1-2, pp. 207–217

Hwang, J.; Lee, B. I.; Klep, V.; Luzinov, I. (2008). Transparent hydrophobic organic-inorganic nanocomposite films. *Materials Research Bulletin,* 43, 10, pp. 2652-2657

Jadav, G. L.; Singh, P. S. (2009). Synthesis of novel silica-polyamide nanocomposite membrane with enhanced properties. *Journal of Membrane Science,* 328, 1-2, pp. 257-267

Jiao, W. M.; Vidal, A.; Papirer, E.; Donnet, J.B. (1989). Modification of Silica Surfaces by Grafting of Alkyl Chains Part III. Particle/Particle Interactions: Rheology of Silica Suspensions in Low Molecular Weight Analogs of Elastomers. *Colloids and Surfaces,* 40, pp. 279-291

Kawakami, H.; Mikawa, M.; Nagaoka, S. (1996). Gas permeability and selectivity through asymmetric polyimide membranes. *Journal of Appled Polymer Science,* 62, 7, pp. 965–971

Kawakami, H.; Mikawa, M.; Nagaoka, S. (1998). Gas transport properties of asymmetric polyimide membrane with an ultrathin surface skin layer. *Macromolecules,* 31, 19, pp. 6636–6638

Kawakami, H.; Nakajima, K.; Shimizu, H.; Nagaoka S. (2003). Gas permeation stability of asymmetric polyimide membrane with thin skin layer: effect of polyimide structure. *Journal of Membrane Science,* 212, 1-2, pp. 195–203

Kikuchi, H.; Yukimon, Y.; Itonaga, S. (2002). Inverter-surge-resistant enameled wire based on nano-composite insulating material. *Hitachi Cable Review,* 21, pp. 55-62

Kim, J. Y.; Mulmi, S.; Lee, C. H.; Park, H. B.; Chung, Y. S.; Lee, Y. M. (2006). Preparation of organic–inorganic nanocomposite membrane using a reactive polymeric dispersant and compatibilizer: Proton and methanol transport with respect to nano-phase separated structure. *Journal of Membrane Science,* 283, 1-2, pp. 172-181

Kim, T. K.; Kang, M.; Choi, Y. S.; Kim, H. K.; Lee, W.; Chang, H.; Seung, D. (2007). Preparation of Nafion-sulfonated clay nanocomposite membrane for direct menthol fuel cells via a film coating process. *Journal of Power Sources,* 165, 1, pp. 1-8

Kusakabe, K.; Ichiki, K.; Hayashi, J.; Maeda, H.; Morooka, S. (1996). Preparation and characterization of silica-polyimide composite membranes coated on porous tubes for CO_2 separation. *Journal of Membrane Science,* 115, 1, pp. 65-75

Lee, Y. L.; Du, Z. C.; Lin, W. X.; Yang, Y. M. (2006). Monolayer behavior of silica particles at air/water interface:A comparison between chemical and physical modifications of surface. *Journal of Colloid Interface Science,* 296, 1, pp. 233-241

Ma, J.; Yang, Z.; Wang, X.; Qu, X.; Liu, J.; Lu, Y.; Hu, Z.; Yang, Z. (2007). Flexible bi-continuous mesostructured inorganic/polymer composite membranes. *Polymer,* 48, 15, pp. 4305-4310

Mosher, B. P.; Wu, C.; Sun, T.; Zeng, T. (2006). Particle-reinforced water-based organic-inorgaic aocomposite coatings for tailored applications. *Journal of Non-Crystalline Solids,* 352, 30-31, pp. 3295-3301

Oh, C.; Lee, Y. G.; Jon, C. U.; Oh, S. G. (2009). Synthesis and characterization of hollow silica microspheres functionalized with magnetic particles using W/O emulsion method. *Colloids and Surfaces A: Physicochemical and Engineering Aspects*, 337, 1-3, pp. 208-212

Ohki, Y. (2005). Study on dielectric properties of LDPE-based nanocomposites by J-power systems. *IEEE Electrical Insulation Magazine*, 21, 3, pp. 55-56

Ouabbas, Y.; Chamayou, A.; Galet, L.; Baron, M.; Thomas, G.; Grosseau, G.; Guilhot, B. (2009). Surface modification of silica particles by dry coating: Characterization and powder aging. *Powder Technology*, 190, 1-2, pp. 200-209

Ranade, A.; Souza, N.; Gnade, B. (2002). Exfoliated and intercalated polyamide-imide nanocomposites with montmorllonite. *Polymer*, 43, 13, pp. 3759-3766

Rangsunvigit, P.; Imsawatgul, P.; Na-ranong, N.; O'Haver, J. H.; Chavadej, S. (2008) Mixed surfactants for silica surface modification by admicellar polymerization using a continuous stirred tank reactor. *Chemical Engineering Journal*, 136, 2-3, pp. 288-294

Rankin, S. E.; Macosko, C. W.; McCormick, A. V. (1998). Sol-Gel Polycondensation Kinetic Modeling: Methylethoxysilanes. *AIChE Journal*, 44, 5, pp. 1141-1156

Rupnowski, P.; Gentz, M.; Kumosa, M. (2006). Mechanical response of a unidirectional graphite fiber /polyimide composite as a function of temperature. *Composites Science and Technology*, 66, 7-8, pp. 1045-1055

Stathatos, E.; Lianos, P.; Tsakiroglou, C. (2004). Highly efficient nanocrystalline titania films made from organic/inorganic nanocomposite gels. *Microporous and Mesoporous Materials*, 75, 3, pp. 255-260

Sun, S.; Li, C.; Zhang L.; Du, H.L.; Burnell-Gray J.S. (2006). Effects of surface modification of fumed silica on interfacial structures and mechanical properties of poly(vinyl chloride) composites. *European Polymer Journal*, 42, 7, pp. 1643-1652

Wu, J.; Yang, S.; Gao, S.; Hu, A.; Liu, J.; Fan, L. (2005). Preparation, morphology and properties of nano-sized Al_2O_3/polyimide hybrid films. *European Polymer Journal*, 41, 1, pp. 73-81

Wu, T.; Ke, Y. (2006). Preparation of silica–PS composite particles and their application in PET. *European Polymer Journal*, 42, 2, pp. 274-285

Xu, J.; Wong, C.P. (2007). Characterization and properties of an organic–inorganic dielectric nanocomposite for embedded decoupling capacitor applications. *Composites Part A: Applied Science and Manufacturing*, 38, 1, pp. 13-19

Xue, L.; Li, J.; Fu, J.; Han, Y. (2009). Super-hydrophobicity of silica nanoparticles modified with vinyl groups. *Colloids and Surfaces A: Physicochemical and Engineering Aspects*, 338, 1, pp. 15-19

Yang, Y.; Wang, P. (2006). Preparation and characterizations of a new PS/TiO_2 hybrid membrane by sol–gel process. *Polymer*, 47, 8, pp. 2683-2688

Yanagishita, H.; Kitamoto, D.; Haraya, K.; Nakane, T.; Okada, T.; Matsuda, H.; Idemoto, Y.; Koura N. (2001). Separation performance of polyimide composite membrane prepared by dip coating process. *Journal of Membrane Science*, 188, 2, pp.165–172

Zhang, C.; Zhang, M.; Cao, H.; Zhang, Z.; Wang, Z.; Gao, L.; Ding, M. (2007). Synthesis and properties of a novel isomeric polyimide/SiO$_2$ hybrid material. *Composites Science and Technology*, 67, 3-4, pp. 380-389

Zheng, P.; Kong L. X.; Li, S.; Yin, C.; Huang M. F. (2007). Self-assembled natural rubber/silica nanocomposites: Its preparation and characterization. *Composites Science and Technology*, 67, 15-16, pp. 3130-3139

Zhong, S.; Li, C.; Xiao, X. (2002). Preparation and characterization of polyimide–silica hybrid membranes on kieselguhr–mullite supports. *Journal of Membrane Science*, 199, 1-3, pp. 53–58

14

Scanning Electron Microscope for Characterising of Micro- and Nanostructured Titanium Surfaces

Areeya Aeimbhu

Department of Physics, Faculty of Science, Srinakharinwirot University, Bangkok, Thailand

1. Introduction

Titanium and its alloys have been used broadly and successfully for numerous applications such as sport equipment [1], aerospace industry [2], marine application [3], medical applications [4] because of its optimum mechanical properties, outstanding corrosion resistance and bio-inert due to the presence of a thin layer of titanium oxide which a naturally formed onto the titanium surface [5-6]. This layer mainly consists of titanium dioxide or titania. Properties of oxide films covering titanium implant surfaces are a key role for a successful osseointegration [7-8], wear and corrosion resistance [9]. Moreover, titania has a large number of potentials in water photoelectrolysis and photocatalysis [10], sensors [11], wastewater remediation [12], automotive industry [13], industrial applications [14] and micro-optoelectronic applications [15]. However, there are some disadvantages of the native titanium oxide which has poor mechanical properties and easily fractured under small scale of fretting and wears conditions [16]. Therefore, many surface modification treatments for examples anodisation (anodic oxidation), cathodic electrodeposition and sol-gel reactions [17-22] have been studied in order to improve the performance of titanium. Moreover, surface properties such as topography, chemical composition and hydrophilicity have an effect on the mechanical stability of the implant-tissue interface. Various surface modification methods of titanium have been shown to improve interfacial interactions at the bone-implant interface and their clinical performance. Moreover, the biological performance of implantable titanium depends crucially on their surface topography in the micrometre (structures larger than 1 micron) and nanometre (structures smaller than 1 micron) range. Surface micro- and nano-topography can reduced inflammatory and guide direct osteoblast responses by altering adhesion, recruitment, movement, morphology, apoptosis and gene expression, and subsequently protein production [23-25]. An anodisation is a simple and an inexpensive technique to prepare thin film titania on titanium surface in different conditions and electrolytes such as acidic, basic, neutral, organic and inorganic which affect surface architecture and chemical composition [26-27]. Anodised oxide layer has thickness in the range of 20 to 180 nm which is thicker than a naturally formed oxide [28]. Moreover, this technique is presently used to achieve micro- and nano-topography surfaces. An anodisation is an electrolytic etching for coating the surface of a metal with an oxide layer which changes the microscopic texture of the surface and the crystal structure of the metal near the surface. An anodisation process accelerates the formation of an oxide coating under controlled conditions to provide the desired result. Since the coating is biocompatible as well as nontoxic, the process lends itself to achieve drastic improvement in implant performance. By adjusting the anodisation condition such as electrolyte, pH, voltage and time, micro- and nano-scale properties could be controlled.

In this article, the morphological of surface was studied by means of scanning electron microscope (SEM) after surface modification in order to evaluate qualitatively the effect of the anodisation conditions. For this purpose, the SEM uses to examine morphological of development of anodised film of titanium that were prepared under different controlled conditions.

2. Materials and methods

2.1 Preparation of surface

Prior to anodisation, commercially pure titanium grade 2 were provided by Prolog Titanium Co., Ltd and cut into 1 cm × 1 cm squares. Titanium sheets were abraded mechanically using silicon carbide (SiC) abrasive papers (Buehler) number 120 to 2000 and rinsed with distilled water. The surface of sheets were sequentially polished to a mirror finished with aqueous alumina (Al_2O_3) 5, 1, 0.3 and 0.05 micron (Buehler, Alpha Micropolish II). Afterward clean with acetone in ultrasonic bath.

2.2 Remove oxide

Titanium surfaces were immersed in an acid mixture (2 ml 48% HF (48% in water) which purchased from Panreac + 3 ml 70% HNO_3 (70% in water) which purchased from Fluka + 100 ml DI water) about 90 seconds to remove the naturally formed oxide layer and then immediately treated by anodisation. All electrolytes were prepared with reagent grade chemicals.

2.3 Anodisation

The electrochemical cell consists of a two electrodes system which graphite acting as the counter electrode (cathode) and titanium sheet acting as the working electrode (anode). The anode and cathode were connected by copper wires and were linked to a positive and negative port of a DC power supply (KMB 3002), respectively. During processing, the anode and cathode were kept parallel with a separation distance of about 1 cm, and were immersed into an electrolyte. Anodisation was performed under potentiostatic conditions at 20 volts. All experiments were carried out at room temperature. Test conditions applied vary in terms of the concentration of acetic acid, the concentration of hydrofluoric acid and an anodisation time. Different series of experiments have been performed by changing the associated experimental parameters as shown in Table 1. After the electrochemical process, the titanium sheets were immediately rinsed by deionised water and heated at 120°C for 30 minutes.

2.4 Surface characterisation

The surface morphology of the samples was observed by using Scanning electron microscope (SEM: JEOL 6300) with Dispersive X-Ray Spectroscopy (EDS) technique.

	Acetic concentration (M)	Hydrofluoric acid (wt%)	Anidising time (minutes)
Series 1	0.001, 0.01, 0.1, 1 and 10M	-	120
Series 2	1 M	-	30, 60, 120, 240, 360 and 480
Series 3	1 M	0.075	30, 60, 360, 480, 720, 1440, 2160 and 2880
Series 4	1 M	0.5, 1.5 and 2.0	60

Table 1. Parameters used for anodisation

3. Result and discussion

In order to find out the effect of anodisation electrolyte and anodisation time on the forming of the oxide at constant voltage in an ambient temperature; the structure of morphology were investigated. Experimental with different anodisation conditions: the concentration of acetic acid, the concentration of hydrofluoric acid and an anodisation time were lead to get an overall view of the formation process of titania nanotubes.

3.1 Effect of concentration of acetic acid on the surface topography

Figure 1 displayed scanning electron microscope (SEM) photographs of an untreated and the anodised titanium surface at 20 volts in different electrolytes concentration. An untreated sample has no visible scratches or wrinkles (figure 1a). The effect of concentration of electrolytes on the surface morphology of the anodic oxide film formed on titanium was monitored. Based on the experimental observation, it was showed that the passive film formed on titanium surface in low concentrations of acetic acid range which from 0.001 to 0.1 M (figure 1(b-d)). The flower-like structures [29] were visible on the substrates which anodised in 1 M acetic acid. Moreover, upon increase concentration of acetic acid the surface morphologies shown the concentration of the flower was developed and increased as shown in figure 1e - 1f. The average diameter of the flowers formed in an acetic acid for 120 minutes was around 300 nanometres.

3.2 Effect of anodisation time on the surface topography

The effect of anodisation time on the morphology of an oxide film in 1M acetic acid was investigated. Figure 2(b-g) summarised the effect of anodisation time on the development of an anodised film in 1M acetic acid at 20 volts. Within 30 minutes (figure 2b), SEM image revealed the formation of a continuous of oxide film over the surface. Flower-like structures were developed at 60 minutes (figure 2c). The density of the flower oxides was observed with the increasing of anodisation time. The average diameter of the flowers increases from a few hundred nanometres to 1200 nanometres.

The effect of hydrofluoric acid and anodisation time on the formation of titania nanotubes in 1M acetic acid with 0.075 wt% HF is studied. SEM image of the surface morphologies obtained in 1 M acetic acid with 0.075 %wt HF at different anodisation time are shown in figure 3(b-i).

The results demonstrated at the initial stage of anodisation (Figure 3(b-c)), the tube-like features appeared over the titanium surface. For longer anodisation time, it can be seen from figure 3(e-f) that the architecture transformation from tube-like nanostructure to a sponge-like titania film with nanoholes. Further anodisation time, the surface of nanoholes is covered with the oxide film (figure 3(g-h)). When the anodisation time is prolonged to 2880 minutes, the discontinuous oxide film covered the titanium surface. The results confirm that fluorine containing electrolyte is a capable electrolyte for anodic formation of titania nanotubes.

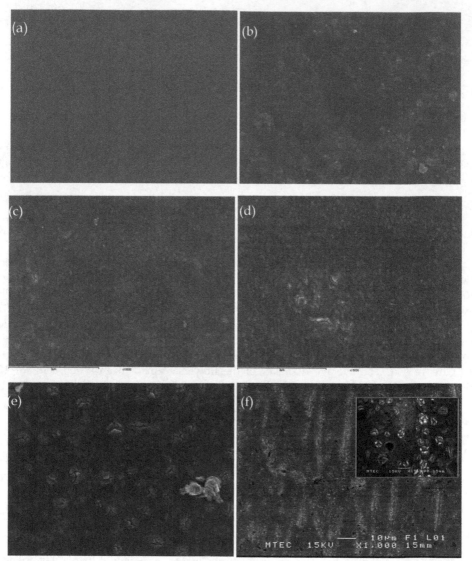

Fig. 1. SEM images of titanium sheets (a) untreated surface and anodised in (b) 0.001 (c) 0.01 (d) 0.1 (e) 1 and (f) 10M acetic acid at 20 volts for 120 minutes at room temperature.

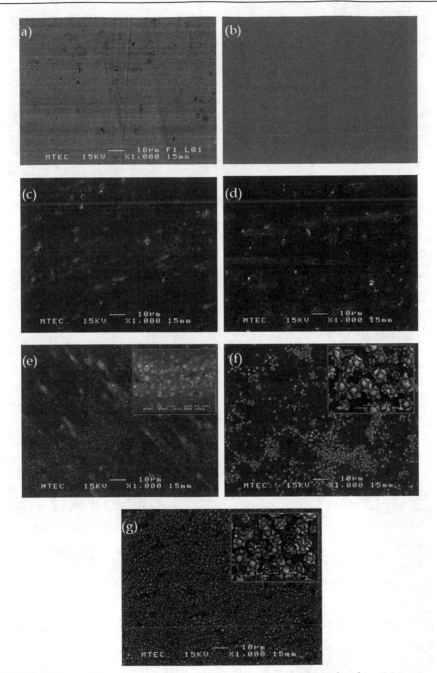

Fig. 2. SEM images of titanium sheets (a) untreated surface and anodised in 1M acetic acid at 20 volts for (b) 30 minutes (c) 60 minutes (d) 120 minutes (e) 240 minutes (f) 360 minutes and (g) 480 minutes at room temperature.

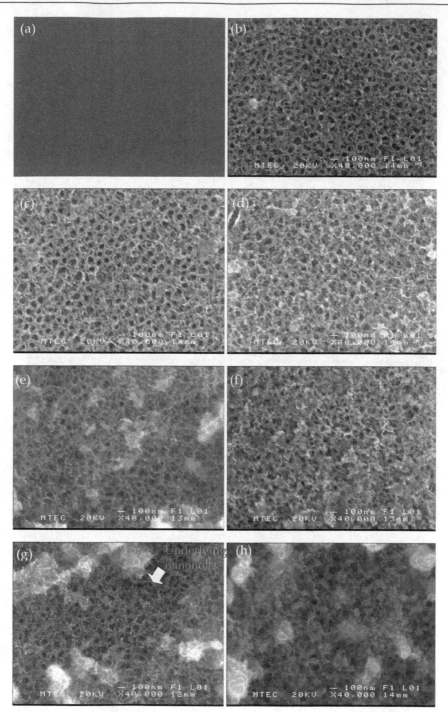

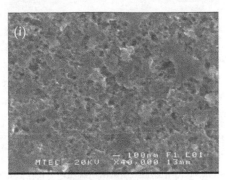

Fig. 3. SEM images of titanium sheets (a) untreated surface and anodised in 1M acetic acid with 0.075 wt% HF at 20 volts for (b) 30 minutes (c) 60 minutes (d) 360 minutes (e) 480 minutes (f) 720 minutes (g) 1440 minutes (h) 2160 minutes and (i) 2880 minutes at room temperature.

3.3 Effect of concentration of hydrofluoric acid on the surface topography

The surface topologies of titanium sheets anodised in electrolyte containing 0.1 M acetic acid with different concentration of HF: 0.075, 0.5, 1.5 and 2.0 wt% are shown in Figure 4. The anodisation was carried out at 20 volts. A network structure appears on the anodised titanium surfaces with concentration of 0.075 wt% HF (Figure 4b). Anodisation in 0.5 wt% and 1.5 wt% HF containing occur a highly ordered and uniform titanium oxide nanotube arrays (Figure 4 (d-e)). The average nanotube inner diameter is approximately 74.5 and 76.5 nanometre, respectively. As the concentration of hydrofluoric concentration was further increased to 2.0 wt%, the surface architecture developed sponge-like (Figure 4f). It points out that the concentration of hydrofluoric acid affect the morphology of titanium surface.

The anodic growth of compact oxides on titanium substrate and the formation of nanotubes in fluoride-containing electrolytes is the result of key processes [30] which are (1) Field-assisted oxidation of the titanium metal that leads to oxide growth at the surface due to interaction of titanium with O^{2-} or OH^- ions. An initial oxide layer formed on the substrate, these anions travel through the oxide layer reaching the titanium/oxide interface. (2) Titanium metal ion (Ti^{4+}) migrate from the substrate at the titanium/oxide interface; Ti^{4+} cations will ejected from the titanium/oxide interface under application of an electric field that move towards the oxide/electrolyte interface. (3) Field-assisted dissolution of titanium metal ion at the oxide/electrolyte interface into the electrolyte. (4) Chemical dissolution of titanium metal ion and TiO_2 due to etching away by fluoride ions. The reactions are [31]:

at anode:

1. oxidation of titanium metal which releases titanium metal ions (Ti^{4+}) and electron

$$2Ti \rightarrow 2Ti^{4+} + 8e^- \qquad (1)$$

2. interaction of titanium metal ions with O^{2-} or OH^- ions

$$Ti^{4+} + 4OH^- \rightarrow Ti(OH)_4 \qquad (2)$$

$$Ti^{4+} + 2O^{2-} \rightarrow TiO_2 \tag{3}$$

Equation (2) and (3) elucidate the hydrated anodic layer and the oxide layer. Further oxide is produced when the hydrated anodic layer releases water by a condensation reaction as the following equation:

$$Ti(OH)_4 \rightarrow TiO_2 + 2\,H_2O \tag{4}$$

at cathode:

$$8H^+ + 8e^- \rightarrow 4H_2 \tag{5}$$

The overall process for anodic oxidation of titanium can be represented as:

$$Ti + 2H_2O \rightarrow TiO_2 + 2H_2 \tag{6}$$

In the presence of fluoride ions electrolyte, fluoride ions enters the $Ti(OH)_4$ or anodic titanium oxide as the following equation:

$$TiO_2 + 6F^- + 4H^+ \rightarrow TiF_6^- + H_2O \tag{7}$$

$$Ti(OH)_4 + 6F^- \rightarrow TiF_6^- + 4OH^- \tag{8}$$

$$Ti^{4+} + 6F^- \rightarrow TiF_6 \tag{9}$$

Equation (7-8) is the mechanism of the pit formation due to the localised dissolution of the oxide and hydrated anodic layer. Then these pits transfer to bigger pores and the pore density increases subsequent to uniformly pores over the titanium surface, with the TiO_2 pores growing more and more deeply into the titanium metal. As described above, the formation of nanotubes govern by a competition between anodic oxidation and chemical dissolution of the oxide as soluble fluoride complexes.

The concentration of the HF added to the electrolyte was varied from 0.075 to 2.0 wt%. As a result, a large change in surface architecture was observed. An average diameter of the nanotubes increased as the fluoride content in electrolyte. As mentioned above, the formation of titania nanotubes determined by the oxide growth rate and the dissolution rate. A result shows that with 0.5 and 1.5 wt% HF, the dissolution rate was slow and resulted in small pore size. With increasing the concentration of HF, the dissolution rate increased and resulted in big pore. It is obvious that the dissolution rate was extremely high with 2 wt% HF because the morphology is not uniform.

The result of the chemical analysis by Energy Dispersive X-ray spectroscopy (EDS) indicated that anodised titanium in 1 M acetic acid with 0.075 wt% HF for 60 minutes at room temperature titania film are shown in figure 5b. The EDS measurements present an oxygen and titanium proportion of 34.914% and 65.086 %, respectively. The result reviewed that the chemical composition of anodic film was nonstoichiometric and the atomic ration of Ti/O is approximately 1.86. The presented of nonstoichiometric structure on the substrate layer implied that some defect exist in TiO_2 nanotubes due to oxygen deficiencies which can cause the formation of crystallographic shear planes and active Ti-sites for the adsorption and chemisorptions of OH groups or other contaminants [32].

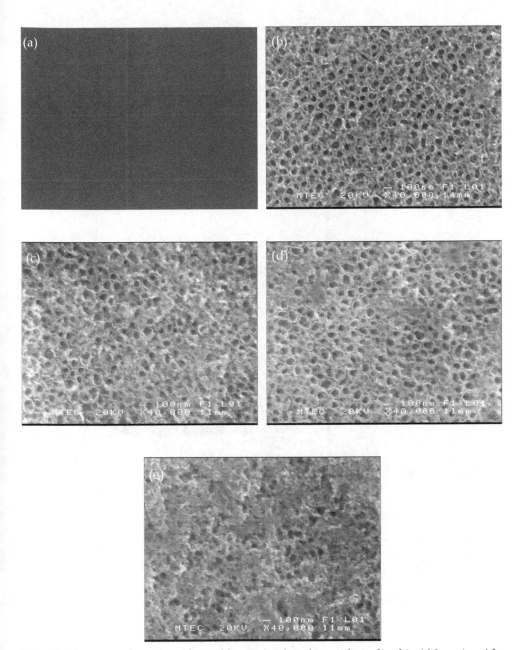

Fig. 4. SEM images of titanium sheets (a) untreated surface and anodised in 1 M acetic acid with (b) 0.075 wt% HF (c) 0.5 wt% HF (d) 1.5 wt% HF (e) 2.0 wt% HF for 60 minutes at room temperature.

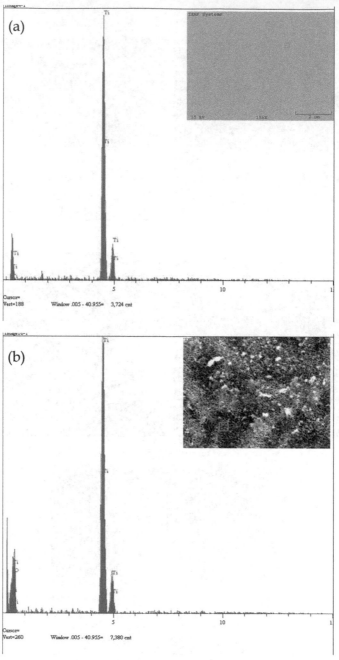

Fig. 5. SEM image of (a) untreated surface and (b) anodised in 1 M acetic acid with 0.075 wt% HF for 60 minutes at room temperature then EDS analysis of the surfaces for energy dispersive analysis was 15 kV, Takeoff Angle 35.0° and Elapsed Livetime 10.0.

4. Conclusion

SEM analysis can clearly showed that anodisation is a simple and economical method to synthesise various surface patterns and textures on the surface of a metallic titanium surface. Moreover, surface morphology is strongly affected by anodisation condition. SEM revealed that the anodisation condition caused micro- and nanomorphological alterations of titanium surface, whereas prolonged exposure to electrolyte resulted in micromorphological changes of the titanium surface. The above result clearly point out that the hydrofluoric acid play important role in controlling the formation of titania nanotubes. Moreover, by adding fluoride ions into the electrolyte, nanotubes can be fabricated under suitable conditions

5. References

[1] M Yamada, *Mats Sci & Eng A.*, 213 (1996) 8.

[2] R R Boyer, *Mats Sci & Eng A.*, 213 (1996) 103.

[3] I V Gorynin, *Mats Sci & Eng A.*, 263 (1999) 112.

[4] M Balazic and J Kopac, *Int. J. Nano and Biomaterials.*, 1 (2007) 3.

[5] M Cortada, L L Giner, S Costa, F J Gil, D Rodríguez, J A Planell, *J of Mats Sci: Mats in Med.*, 11 (2000) 287.

[6] A W E Hodgson, Y Mueller, D Forster, S Virtanen, *Electrochemical Acta.*, 47 (2002) 1913.

[7] T Albrektsson and B Albrektsson, *Acta Orthopaedica.*, 58 (1987) 567.

[8] D Leonardis, A K Garg, G E Pecora, *Int. J. of Oral & Maxillofacial Implants.*, 14 (1999) 1.

[9] Y X Leng, J Y Chen, P Yang, H Sun, N Huang, *Surf & Coating Technol.*, 166 (2003) 176.

[10] D M Blake, P C Maness, Z Huang, E J Wolfrum, J Huang, *Separation and Purification Methods.*, 28 (1999) 1.

[11] J A Byrne, J W J Hamilton, T A McMurray, P S M Dunlop, V J A Donaldson, J Rankin, G Dale, D Al Rousan. *Clean Technol.*, (2007) 242.

[12] A S Stasinakis, *Global NEST J.*, 10 (2008) 376.

[13] Y Yamashita, I Takayama, H Fujii, T Yamazaki, *Nippon Steel Technical Report.*, 85 (2002) 11.

[14] G F New, *Current Science.*, 3 (1973) 133.

[15] R Vogel, P Meredith, I Kartini, M Harvey, J D Riches, A Bishop, N Heckenberg, M Trau, H Rubinsztein-Dunlop, *ChemPhysChem.*, 4 (2003) 595.

[16] S A Brown and J E Lemons, *ASTM STP 1272, ASTM.* (Philadelphia, 1996).

[17] X Liu, P K Chu, C Ding, *Mats Sci & Eng R.*, 47 (2004) 49.

[18] J M Lee, Y S Kim, C W Kim, K S Jang, Y J Lim, *J Korean Acad Prosthodont.*, 42 (2004) 352.

[19] D P Dowling, P V Kola, K Donnelly, T C Kelly, K Brumitt, L Lloyd, R Eloy, M Therin, N Weill, *Diamond and Related Mats.*, 6 (1997) 390.

[20] C C Chen, J H Chen, C G Chao, *J of Mats Sci.*, 40 (2005) 4053.

[21] A M PeirÓ, E Brillas, J Peral, X Domènech, J A AyllÓn, *J of Mats Chem.*, 12 (2002) 2769.

[22] Y Li, J Hagen, W Schaffrath, P Otschi, D Haarer, *Solar Energy Material and Solar Cells.*, 56 (1999) 167.

[23] S Lavanus, G Louarn, P Layrolle, *Int. J. of Biomaterials.*, 2010 (2010) 1-9.

[24] F Variola, F Vetrone, L Richert, P Jedrzejowski, J.-H Yi, S Zalzal, S Clair, A Sarkissian, D.F Perepichka, J.D Wuest, F Rosei, A Nanci, *Small.*, 5 (2009) 996-1006.

[25] C Toth, G Szabó, L Kovács, K Vargha, J Barabás, Z Németh, *Smart Mater. Struct.*, 11 (2002) 813-818.

[26] T Shibata and Y C Zhu, *Corrosion Science.*, 37 (1995) 253.

[27] K H Kim and N Ramaswamy, *Dental Mats J.*, 28 (2009) 20.

[28] Metalast Technical Bulletin. (NV: Metalast International Inc, 2000).

[29] L Bartlett, *Optics & Laser Technol.*, 38 (2006) 440.

[30] K G Mor, O K Varghese, M Paulose, K Shankar, C A Grimes, *Solar Energy Materials & Solar Cells.*, 90 (2006) 2011.

[31] A Jaroenworaluck, D Regonini, C R Bowen, R Stevens, D Allsopp, J Mater Sci., 42 (2007) 6729-6734.

[32] M K Kyung, S J Doo, S H Cheol, *Nanotechnology.*, 22 (2011) 1-17.

Permissions

The contributors of this book come from diverse backgrounds, making this book a truly international effort. This book will bring forth new frontiers with its revolutionizing research information and detailed analysis of the nascent developments around the world.

We would like to thank Viacheslav Kazmiruk, for lending his expertise to make the book truly unique. He has played a crucial role in the development of this book. Without his invaluable contribution this book wouldn't have been possible. He has made vital efforts to compile up to date information on the varied aspects of this subject to make this book a valuable addition to the collection of many professionals and students.

This book was conceptualized with the vision of imparting up-to-date information and advanced data in this field. To ensure the same, a matchless editorial board was set up. Every individual on the board went through rigorous rounds of assessment to prove their worth. After which they invested a large part of their time researching and compiling the most relevant data for our readers. Conferences and sessions were held from time to time between the editorial board and the contributing authors to present the data in the most comprehensible form. The editorial team has worked tirelessly to provide valuable and valid information to help people across the globe.

Every chapter published in this book has been scrutinized by our experts. Their significance has been extensively debated. The topics covered herein carry significant findings which will fuel the growth of the discipline. They may even be implemented as practical applications or may be referred to as a beginning point for another development. Chapters in this book were first published by InTech; hereby published with permission under the Creative Commons Attribution License or equivalent.

The editorial board has been involved in producing this book since its inception. They have spent rigorous hours researching and exploring the diverse topics which have resulted in the successful publishing of this book. They have passed on their knowledge of decades through this book. To expedite this challenging task, the publisher supported the team at every step. A small team of assistant editors was also appointed to further simplify the editing procedure and attain best results for the readers.

Our editorial team has been hand-picked from every corner of the world. Their multi-ethnicity adds dynamic inputs to the discussions which result in innovative outcomes. These outcomes are then further discussed with the researchers and contributors who give their valuable feedback and opinion regarding the same. The feedback is then collaborated with the researches and they are edited in a comprehensive manner to aid the understanding of the subject.

Apart from the editorial board, the designing team has also invested a significant amount of their time in understanding the subject and creating the most relevant covers. They scrutinized every image to scout for the most suitable representation of the subject and create an appropriate cover for the book.

The publishing team has been involved in this book since its early stages. They were actively engaged in every process, be it collecting the data, connecting with the contributors or procuring relevant information. The team has been an ardent support to the editorial, designing and production team. Their endless efforts to recruit the best for this project, has resulted in the accomplishment of this book. They are a veteran in the field of academics and their pool of knowledge is as vast as their experience in printing. Their expertise and guidance has proved useful at every step. Their uncompromising quality standards have made this book an exceptional effort. Their encouragement from time to time has been an inspiration for everyone.

The publisher and the editorial board hope that this book will prove to be a valuable piece of knowledge for researchers, students, practitioners and scholars across the globe.

List of Contributors

Otávio da Fonseca Martins Gomes and Sidnei Paciornik
Centre for Mineral Technology – CETEM, Dept. of Materials Engineering, Catholic University of Rio de Janeiro, Brazil

Junsuke Fujiwara
Osaka University, Japan

Maribel L. Saucedo-Muñoz, Victor M. Lopez-Hirata and Hector J. Dorantes-Rosales
Instituto Politecnico Nacional (ESIQIE), Mexico

Yoshio Ichida
Utsunomiya University, Japan

Suksun Horpibulsuk
Suranaree University of Technology, Thailand

Eva Tillová, Mária Chalupová and Lenka Hurtalová
University of Žilina, Slovak Republic

Thomas N. Otto, Wilhelm Habicht, Eckhard Dinjus and Michael Zimmermann
Karlsruhe Institute of Technology, IKFT, Germany

Osei-Wusu Achaw
Department of Chemical Engineering, Kumasi Polytechnic, Kumasi, Ghana

Laura Frisk
Tampere University of Technology, Department of Electronics, Finland

A. Alyamani
National Nanotechnology Research Centre, KACST, Riyadh, Saudi Arabia

O. M. Lemine
Physics Department, College of Sciences, Imam University, Riyadh, Saudi Arabia

Mukul Dubey and Hongshan He
Center for Advanced Photovoltaics, Department of Electrical Engineering & Computer Science South Dakota State University, Brookings, SD, USA

Devina Sharma, H. Kishan and V.P.S. Awana
Quantum Phenomena and Applications, National Physical Laboratory (CSIR), New Delhi, India

M. Husain
Department of Physics, Jamia Millia Islamia, New Delhi, India

Ranjan Kumar
Department of Physics, Punjab University, Chandigarh, India

Shiva Kumar Singh
Quantum Phenomena and Applications, National Physical Laboratory (CSIR), New Delhi, India
Department of Physics, Jamia Millia Islamia, New Delhi, India

Xiaokun Ma and Sun-Jae Kim
Institute/Faculty of Nanotechnology and Adv. Materials Engin., Sejong University #98 Gunja-dong, Gwangjin-gu, Seoul, South Korea

Areeya Aeimbhu
Department of Physics, Faculty of Science, Srinakharinwirot University, Bangkok, Thailand

Printed in the USA
CPSIA information can be obtained
at www.ICGtesting.com
JSHW011456221024
72173JS00005B/1104

9 781632 384065